Computer Games and Simulation for Biochemical Engineering

Computer Games and Simulation for Biochemical Engineering

HENRY R. BUNGAY, III

Rensselaer Polytechnic Institute
Troy, New York

A Wiley-Interscience Publication

JOHN WILEY & SONS
New York · Chichester · Brisbane · Toronto · Singapore

Copyright ©1985 by John Wiley & Sons, Inc.

All rights reserved. Published simultaneously in Canada.

Reproduction or translation of any part of this work beyond that permitted by Section 107 or 108 of the 1976 United States Copyright Act without the permission of the copyright owner is unlawful. Requests for permission or further information should be addressed to the Permissions Department, John Wiley & Sons, Inc.

Library of Congress Cataloging in Publication Data:

Bungay, Henry Robert, 1928-
   Computer games and simulation for biochemical engineering.
   "A Wiley-Interscience publication."
   Includes bibliographies and index.
   1. Biochemical engineering—Data processing.
2. Biochemical engineering—Mathematical models.
I. Title. TP248.3.B86 1985   660'.63'071   84-19510
ISBN: 0-471-81278-1

Printed in the United States of America

10 9 8 7 6 5 4 3 2 1

PREFACE

The two topics of this book, simulation and teaching games, have much in common. Games usually employ simulation, and there is a component of learning in all simulation even if the inventor is the only participant. Constructing a simulation requires knowledge of the system, but the absence of solid information leads to guesses about its structure. The insight gained while attempting simulation often has more value than the simulation itself. This author is disappointed when someone develops an elegant simulation, publishes, and moves on expecting that others will use it. Seldom will others verify the simulation by comparison with actual data and use it for prediction or for testing ideas. Simulation is more likely to be put to good use when there is a recognized need for assistance with the computer for a particular problem, and the inventor is intimately associated with the overall project. Ideas can be tested quickly with the computer, and experimental results can be compared to the simulation. This leads to an organized approach to explanations and interpretations of data that are not at all obvious to someone that does not have skill in simulation.

At Rensselaer Polytechnic Institute, we have cast a number of computer simulations into the form of teaching games that are effective aids to our courses. For over twelve years, games have been used with great success in a first-year graduate course, Biochemical Engineering. A new summer course, Advanced Biochemical Engineering, starts out with computer simulation techniques to provide tools for studying other topics and has some additional sophisticated teaching games. Students are given the computer tools right at the start of our courses and are encouraged to use them. It has been surprising and gratifying to find that all of our summer attendees so far have had familiarity with computers. They have been able to start very quickly on computer exercises with almost no lag time for learning how to handle passwords, account numbers, computer notation, and the like.

Simulation is so much a part of the RPI program that inclusion in biochemical engineering seems natural. A comprehensive look at many ways that computer simulation can be used in biochemical engineering should enlarge the perspectives for this field and encourage others to develop the necessary skills.

This author was performing laboratory research in conjunction with computer simulation for several years prior to developing an interest in computers for teaching. Teaching games became an interest because of a six-week summer workshop entitled "Simulation Games for Teaching Natural Resource Management" at the University of Washington in 1970. A great deal of luck was involved because the author's very first game was a winner. Since then the success ratio has been acceptable, but all may have been lost if one of the subsequent failures had been the first game and discouraged further efforts.

Some explanation of the use of SIMBAS and SIM4 in this book is needed. There are many fine simulation packages for differential equations available at computer centers all over the world. The author first learned an early program, PACTOLUS, and graduated to CSMP which is very powerful. MIDAS, MIMIC, and GPSS have been tried briefly, but our own languages have been found easy to use and require relatively little computer time. The origin of these programs was a simulation method for minicomputers written by D.C. Stanzione at Clemson University. SIMBAS, a version in BASIC (SIMulation in BASIC) runs very nicely on inexpensive microcomputers and can be taught to a neophyte in less than one-half hour. This has evolved to a FORTRAN translation (SIMulation in 4TRAN), SIM4, that has graphics features. These are "bare bones" programs that have little more than input-output routines and an integration scheme. The graphics routines in SIM4 programs are specific to the RPI system using PRIME computers, but it should not be too difficult to substitute graphics subroutines for other systems.

The feature of our simulation programs that makes learning so easy is a built in example. The student is given a working program that can be modified to handle a new problem. If the changes are made stepwise, it is instructive to compare with the original and simple to diagnose what changes result in troubles.

Some people will consider the use of an elementary language, BASIC, rather peculiar for dealing with fairly sophisticated topics in biochemical engineering. The justification is the ease with which BASIC is understood. Furthermore, general principles are more important than efficient use of the computer, and programs in BASIC can be translated into better languages when this is worthwhile. An overriding consideration is the universality of BASIC that makes it possible to execute these programs on almost any computer. In fact, most of the programs in this book will run on inexpensive personal computers.

Simulation that produces columns of numbers is hard to interpret. We feel that using computer graphics for teaching is much more effective. The technology for computer graphics is here, but there is a reluctance by professors to embark on programs for their courses. They fear that great amounts of time will be consumed in developing effective teaching aids, and the return may not be worth the trouble. This may be true of very elaborate computer-assisted instruction, but there are many simple computer packages that are well-suited to some teaching situations. Computer games, in particular, can be easy to program, and fun in playing them is strong motivation for student involvement.

An awareness of computer teaching games and keeping alert over a period of a dozen years for opportunities to develop them has led the author to ten games used regularly and effectively in courses. Often a new game is written quickly in BASIC and tested with a group of students. All of the games have been translated to faster languages for use at the university computer centers so that the author's microcomputer is not tied up by students. Shortcomings are corrected, and games are improved as experience leads to new ideas. A valuable source of help has been from students who voluntarily improve programs or from those who accept a game as a class assignment. Typical programming time has been roughly 20 hours, although one student took an elementary game of the author's and spent several hundred hours in developing a masterpiece. This is described in Chapter 4 as the MULTI-STAGE CONTINUOUS CULTURE GAME. We have learned a great deal about stages of fermentation, supplemental feeding, and recycle by tinkering with this game.

Except for support to attend the course in 1970, none of the author's work on teaching games has been funded by a government agency. This has been one of the author's hobbies on a very occasional and part-time basis. Our games have started with flashes of inspiration. Although a few of the ideas led to poor games, appreciation of why a game failed helps in designing good games. The examples in this book may stimulate more activity for this form of teaching and others may encounter our situation where students want to play the games far beyond the allotted times. Of the various techniques available for education, simulation assignments and teaching games are particularly effective, deeply involve the students, and are relatively inexpensive.

CONTENTS

Chapter 1	INTRODUCTION	1
	Computer Teaching Games	2
	Computer Graphics	3
	Derivatives	4
	Formulating Differential Equations	6
Chapter 2	SIMULATION TECHNIQUES	9
	SIMBAS Instructions	10
	Sources of Error	13
	Oxygen Dynamics in a Fermenter	15
	Variables That Should Not Become Negative	17
	Two Tanks in Series	17
	Respiration with Oxygen Limitation	19
	Summary	21
	Suggested Practice Problems	24
Chapter 3	BATCH PROCESSES	25
	FERMT, a Fermentation Process Development Game	26
	JERMFERM, a Fermentation Process Game	29
	Synchronous Culture	29
	Mathematics of Batch Culture	31
	Structured Models	33
	Fed Batch Systems	35
	Semi-Continuous Culture	35
	References	37

Chapter 4	CONTINUOUS CULTURE	39
	Mathematical Analysis	40
	MONOD, a Single-Stage Chemostat Game	42
	Recycle	45
	Multi-Stage Continuous Culture	48
	Description of the Program	48
	Execution of the Program	51
	References	55
Chapter 5	MIXED CULTURES	57
	Competition	57
	Concentration-Coupled Continuous Culture	62
	Innoculation of a Slower Culture to a Chemostat	62
	Two-Culture Chemostat with Inoculation	63
	Prey-Predator Systems	67
	Improved Models for Predation	67
	First-Order Delay	77
	Ecosystem Theory	78
	References	79
Chapter 6	ENVIRONMENTAL SYSTEMS	81
	The Dissolved Oxygen Sag Curve	81
	Stream Population Dynamics Game	84
	Modeling of Lakes	85
	Modeling of Waste Treatment	91
	Conclusion	91
	References on Waste Treatment	92
Chapter 7	SELECTED TOPICS	95
	STERIL: Game for Fermenter Sterilization	95
	Setting the Stage for the Sterilization Game	98
	CHROMO, a Game for Teaching Biochemical Processing	101
	Setting the Stage for the Chromo Game	102
	Description of the Program	103
	Matrix Techniques for Modeling Microbial Slime	110
	Mass Balances for Washing and Extraction	112
	Membrane Founling	115
	References	120

		xi
Chapter 8	DYNAMIC ANALYSIS	121
	Transient States	124
	Forcing Functions	125
	Bode Plots	133
	Linearity	135
	Open and Closed Loop Responses	135
	Programs for Dynamic Analysis	138
	Example of Using ICPALS	139
	References	148
	General References for Chemical Process Control	148
Chapter 9	OTHER USES OF COMPUTERS	149
	Computers for Chemistry	149
	Computer Aids for Teaching Microbiology	151
	Computer Control	152
	Mass and Energy Balances	153
	Other Computer Programs for Chemical Engineers	154
	Computer Teaching Games	155
	Drafting and Lettering	155
	Miscellaneous	157
	Conclusion	158
	References	159
Appendix 1	Listing of SIM4	161
Appendix 2	FORTRAN Listing of FERMT	165
	BASIC Listing of FERMT	167
Appendix 3	Listing of MONOD Game	169
Appendix 4	BASIC Listing for Recycle Program	171
	Early Version of Two-Stage Fermentation	172
Appendix 5	BASIC Program for Continuously Inoculated Fermenter	173
	FORTRAN Listing for Continuously Inoculated Fermenter	174
Appendix 6	Listing for STERIL Game	177
Appendix 7	Listing for CHROMO Game	179

Chapter 1

INTRODUCTION

Computer simulation is a means for describing the behavior of systems through the solution of equations. Exercising a simulation by testing many parameters allows great insight, and computer experimentation is several orders of magnitude faster than laboratory experimentation. For example, biological processes for anaerobic decomposition of wastes are very delicate. It sometimes takes several months to get an anaerobic system into balance so that a program of testing variables can begin. Computer simulation of anaerobic digestion is difficult relative to other simulation problems, but when the computer program is ready, variables can be tested in many permutations per day.

Many people, including this author, are highly suspicious of the results of computer simulation. The reasons are that models may be oversimplified or may be so cluttered with extraneous detail that selecting coefficients is a guessing game. It is very dangerous to accept computer results at face value without checking the assumptions or seeking the author's bias. Nevertheless, there is tremendous value in the construction of computer models. The first step is organizing information so that equations can be formulated. This usually forces one to think differently about a biological system and to develop a critical outlook. As the equations are derived, decisions must be made about the terms and effects that are significant. There are almost always processes about which too little is known, or the rates and other coefficients have not been measured. One overriding value to attempting simulation is defining the areas of ignorance. This also suggests new research topics to fill in the gaps of knowledge.

When simulations are compared to actual data, there may be deep satisfactions due to excellent agreement. This is very rare compared to the frequent frustrations resulting from poor to terrible concurrence. Either the data or the equations are wrong, and science and engineering can advance as better lab experiments are performed or as more correct relationships are fed to the computer.

Not only is simulation a learning experience for those constructing the computer program, but others may benefit. The logic of the program, the relationships, the decisions about what factors to include, and the trade-offs of accuracy versus computer cost may interest others who have similar problems. Students can learn from simulation, but there usually will have to be changes in the program to guide them and provisions to avoid traps created by feeding improper information or specifications as they design their computer runs. Taking this a step further, the simulation can be the basis of a game that maximizes teaching value and avoids traps.

COMPUTER TEACHING GAMES

The Rensselaer Polytechnic Institute's Center for Interactive Computer Graphics has two PRIME computers with approximately 40 terminals in addition to a number of separate systems for color graphics. Computer games have become an important part of the biochemical engineering activity at RPI Topics in continuous cultivation, optimization of fermentation, sterilization, and product recovery are well-suited to teaching with computers. As the techniques of interactive computer graphics advance, computer teaching games become more interesting and effective. Although our games were devised for teaching, there have been additional benefits in terms of valuable practical insights and new understanding of complicated systems. Immediate display of results after changing the specifications to the program allows rapid progress in the analysis of a multiplicity of parameter combinations.

Games haves great value as supplements to lectures or as independent self-study modules. The coefficients in the equations take on added meaning when the effects of various combinations can be displayed easily and graphically. For example, a single-stage chemostat game used for over ten years has been evaluated thoroughly by students in biochemical engineering courses. It is effective in solidifying concepts that were introduced by several lectures. There is an ideal fit of interactive computer graphics with continuous culture in rapid analysis of complicated arrangements. With our multi-stage continuous culture game in particular, some of the results were expected and in accord with intuition while other graphs were novel and unexpected. New research areas have been uncovered by noting some interesting configurations of feed and recycle streams in the computer simulation. Furthermore, the introduction to recycle and staged cultures in the lectures and in the text book seems to be expanded to a deep and thorough appreciation by about one hour of interaction with the computer.

An alternative to tinkering with the computer games by each student would be to prepare and hand out samples of computer results that illustrate important points about multistage continuous culture. In our opinion, this is much inferior to assigning the game to students for testing of various options. Individual sessions at the computer require a much greater level of involvement, and the students can pursue an original line and be rewarded by unexpected results. There are creativity and enthusiasms not present in a packaged presentation.

COMPUTER GRAPHICS

Very sophisticated graphics are an everyday affair with arcade games such as PACKMAN, DONKEY KONG, SPACE INVADERS, and with home computers, special effects in movies and television, and computer-drawn art. Although many hours are required to become expert in three-dimensional, high-resolution color graphics, children can produce amazing low-resolution results with their personal computers. At RPI all freshman engineering students are trained in elementary computer graphics, and some take courses in which very advanced techniques are used. Nevertheless, very few people must encode the instructions for fundamental graphics, and most use programs or subroutines designed to make graphics relatively simple. The graphics routines in the programs listed in the appendix are not explained in detail, but comments tell how to change labels, scale factors, and the like.

Graphics subroutines for most computers can draw at specific locations on the display screen or can connect two locations with a straight line. There are special routines to draw boxes or circles. For three-dimensional drawings, there are algorithms for blanking out structures that would be hidden by outer walls if this is desired. There are also algorithms for showing an object rotated about a given axis. Colors and intensities can be specified for lines and areas on color systems.

Only two dimensional graphs are used in this book. Certainly, color would add elegance and clarity to the figures but at a major increase in cost and complexity.

More complicated games may employ features that help the player. Specifications for the program are often entered by typing at the terminal, but light pens that touch areas of the screen can be easy to use. Often a menu of options is displayed, and touching a selection is faster than typing. There are also ways to interact with the display by using the light pen to create boxes, blocks, or figures and to indicate how lines should interconnect them. This is a feature of several programs to be discussed later. The light pen can also "drag" pictures or lines around the display screen. For example, poles or zeros that relate to differential equations can be positioned to the desired location on the screen, and the program will calculate backwards to change the coefficients in the equations to agree with the selected values. Menus and light pens also can be programmed to change scales so that the display fits the screen space better or the plot can be enlarged to show the regions of greatest interest. Another menu option that we sometimes use will save a graph and allow it to be recalled later for comparisons with other graphs.

DERIVATIVES

In most research with physical and biological systems, various data are graphed versus time. Developing relationships and explaining the observed behavior are the crucial points of the research, and mathematical expressions help to organize information. It may be very difficult to derive rational equations for the time behavior because of accumulation of the effects of several factors. However, it is often possible to formulate equations for the rate of change. For example, several microbial cultures may interact in a very complicated fashion such that explaining population fluctuations is impossible. However, the growth rates can be derived from nutrient concentrations and interaction terms based on inhibitions, stimulations, or antagonisms.

Simple differential equations can often be solved explicitly by mathematical methods, but numerical solutions may be needed for non-linear or complicated equations. The general ideas behind numerical techniques are straight forward. If we know the value of a variable at a particular time and can substitute into the differential equation to calculate its instantaneous rate of change, we can project to its new value at a slightly greater time. In other words, given a starting point and the direction of movement, the next point can be estimated. By moving to new times and repeating the process over and over, it is possible to construct the behavior of the property from the equation for its derivative.

Packages for simulation of differential equations have time control features. The user can specify the integration interval (step size), the length of time for the solution, and the interval for output of the results. The independent variable is usually time, but some other variable may be selected. In any case, the units are assigned arbitrarily. One computer unit of time can be one microsecond, one second, one hour, one day, or one century depending on the particular problem. When the integration interval is small, it would be impractical to print the results after each calculation. For example, with an integration of 0.01 time units, assume that a total time of 10 time units is appropriate. If the print interval were equal to the integration interval, there would be one thousand lines of output. Obviously, an output interval of 0.5 or 1.0 would avoid producing many pages of output and would probably provide sufficient information about the solution. The total time for the simulation run again depends on the particular problem. With job specifications to a computer center, it is essential to specify the execution time. But with microcomputers, it may be more convenient to observe the output line-by-line until a decision is reached that the run has proceeded long enough.

If very small increments of time are selected for stepping through the calculations with differential equations, the solution can be extremely accurate. However, the calculations may consume inordinate amounts of computer time and waste money. On the other hand, large steps may develop poor solutions especially when the derivative is changing rapidly. It is recommended when starting a simulation of differential equations that a step size be selected that is likely to be too large for obtaining accurate results. The computer run can be repeated with a smaller step size to see if the results are much different. Its takes relatively little computer money to repeat this test until the answers are changing only by a very small percentage from the previous trial. Furthermore, it may be wise to select a step size that gives only approximate answers while getting the simulation to work and while adjusting coefficients. When the rough work is completed, final runs can employ a small integration step to get elegant answers.

A particularly impressive advance is the development of languages to solve differential equations by simulation. Simulation languages make it easy to solve exceedingly difficult and complicated simultaneous equations. Computer methods for solution require very little knowledge of mathematics. In fact, some equations that greatly challenge the mathematicians are pushovers for computer simulation. The great danger in using the computer methods is violation of some basic rule of mathematics such as inadvertently dividing by zero or integrating through a discontinuity. These dangers are usually unimportant for practical problems particularly in biology or for physical systems when their equations are well behaved, continuous functions. There is no infinity in a real engineering system, and common sense alerts us to troubles when the computer produces impossible results. Thus, except for remote hazards, the simulation languages allow a neophyte to achieve solutions that could tax or even exceed the abilities of a highly trained mathematician.

One more point before getting down to business: computer simulation is not the same thing as conventional computer programming. Languages such as FORTRAN are used to write programs for an infinite variety of problems. It is even possible and fairly common to write programs in these languages for solving differential equations.

FORMULATING DIFFERENTIAL EQUATIONS

The differential equations needed for bioprocesses arise in several ways - kinetic expressions, laws of physics, and mass or energy balances. Let us first consider some reaction kinetics. Order of a reaction is defined as the sum of exponents in its mass action equation. For example:

Zero order: $dA/dt = $ constant

First order: $dA/dt = k\,[A]$

Second order: $dA/dt = k\,[A]^2$ or $dA/dt = k\,[A]\,[B]$

Third order: $dA/dt = k\,[A]^3$ or $=$ 3 terms multiplied

Fractional order: $dA/dt = k\,[A]^{\frac{1}{2}}$

Writing a kinetic expression does not mean that the mechanism of the reaction is defined. In many biological reactions, water is a reactant. However, the systems are aqueous, and the concentration of water remains very nearly constant during the reaction and is not apparent in the kinetic dependencies. If the same reaction were carried out in alcohol, a different kinetic expression would be written including the effect of water. Most zero order reaction arise in situations where a key effect is not considered. One such possibility is a reaction catalyzed by light. The rate may depend upon the intensity of illumination and be independent of the concentrations of reactants.

The example built into the SIMBAS program comes from the reactions:

$$A \longrightarrow B \longrightarrow C$$

with the equations:

$$dA/dt = -k\,[A] \qquad dB/dt = k\,[A] - m\,[B]$$

The concentration of A can only decline, but B rises to a peak as it is made from A and then declines as its further reaction overshaddows its formation because A is disappearing.

A simple guide prevents omitting terms in differential equations based on kinetics. There is one arrow to or from A in the above equation, so there is but one term in the differential equation. For B, there are two arrows and thus two terms. Consider the more complicated scheme:

$$A \longrightarrow B \rightleftharpoons C \longrightarrow D$$
$$\updownarrow \qquad \downarrow$$
$$E \qquad F \longrightarrow G$$

There are reversible reactions and branching pathways. The equation for A will have one term (one arrow), B will have 5 terms (5 arrows), C has four terms, D has one, E has two, and F has two. By the way, E will not be present at extended time because the reaction to form E is reversible and paths to non-reversible products will eventually capture all the intermediates.

Differential equations based on physics come from the laws of force, motion, and the like. For example, the derivative of position is velocity, and the derivative of velocity is acceleration. The physics of structures involves shear and bending moments, and civil engineers often deal with the fourth derivative.

Mass balances are a very common way of analyzing reactor systems. The basic equation is:

rate of change = input rate - output rate + reaction rate

Let us use microbial growth in a vessel as an example of performing a mass balance. Assuming aseptic technique, the input of organisms to the vessel is zero. Feeding sterile nutrient solution will wash out organisms at a rate equal to the flow rate times the concentration of organisms in the vessel. This assumes good mixing such that the exit stream is representative of the contents of any region of the vessel. Growth will occur at some rate dependent on nutrient concentration. There may be death due to organisms growing old. The equation with all these terms is:

change = growth - death - washout

$$V\, dx/dt = \mu\, x\, V - K_d\, V\, x - F\, x$$

where   V = volume of vessel
 x = organism concentration
 μ = specific growth rate coefficient
 F = feed rate for substrate
 Kd = death rate coefficient
 t = time

It is a good idea to always check units and dimensions. These rate of change terms come out in mass/volume,time. The input rate is zero for organisms if the feed is sterile, and the output is mass/volume,time. The increase term is based on the fact that microorganisms grow or divide at a rate proportional to their concentration, and the dimensions for this term are mass/volume,time. There are simultaneous equations for substrate and for cell concentration, but the substrate equation has an input term. Consumption of substrate is related to growth by the coefficient Y. Although it appears to be a routine matter to derive mass balance equations, this is a powerful tool for engineers that is also basic to the analysis of systems in biochemical engineering.

This chapter has been rather elementary, but it is important to be versed in the fundamentals before embarking on other topics. The following chapter is also easy but has specific instructions for performing simulation.

Chapter 2

SIMULATION TECHNIQUES

Two simulation programs will be featured in this book. Examples and simple explanations will emphasize SIMBAS, a BASIC program well suited to personal microcomputers. Much better output is obtained at the R.P.I. Center for Interactive Computer Graphics with SIM4, a FORTRAN translation of SIMBAS but with sophisticated graphics subroutines. The author often uses SIMBAS with a computer in his office for early trials of a new simulation and switches to SIM4 to finish the problem and to get graphs for easy comparison of different runs. Most of the graphs in this book were produced at the graphics center.

Directions are provided for SIMBAS, but they are nearly identical with those for SIM4. A listing of SIM4 in Appendix 1 shows that sections for specifying parameters and writing equations are in about the same relative locations in the two programs.

SIMBAS is programmed in BASIC to solve simultaneous, ordinary differential equations that can be highly non-linear and do not have to be first-order. The integration scheme is a 4th-order Runge-Kutta. The independent variable (usually time) is T. Names beginning with T and followed by an integer are reserved for the program and may not be used in the differential equations, e.g. T1, T2. Standard BASIC operations as shown in Table 2-1 are available as are the functions in Table 2-2. Parentheses must be used with the functions and are commonly used to group portions of an equation together. All of the operations inside the parentheses are completed and the result is used as an entity.

TABLE 2-1

Arithmetic Operations

Symbol	Example	Meaning
+	A + B	Add B to A
-	A - B	Subtract B from A
*	A * B	Multiply A and B
/	A / B	Divide A by B
↑	A ↑ B	Raise A to B power
=	A = A+1	Increase A by one

TABLE 2-2

Functions

SIN(x)	Sine
COS(x)	Cosine
ART(x)	Arctangent
EXP(x)	Exponentiation, raise e to x power
LOG(x)	Natural logarithm
ABS(x)	Absolute value
SQR(x)	Square root

SIMBAS Instructions

SIMBAS is set up with an example of two simultaneous equations. See Figure 2-1. After loading SIMBAS, execute the example by typing RUN (then strike the return key). For new equations, use the example as a guide and replace with the desired equations and values as follows:

1. equations: one per line starting at Line 1000
2. coefficients: use Lines 20 through 55
3. initial conditions: use Lines in the 80's
4. specify N, the number of differential equations, Line 80.
5. specify T1, the integration interval, Line 56
6. specify T2, the total time, Line 57
7. specify T3, the print interval, Line 58
8. output, see Line 52 for headings and Line 2000 for variables.

A line is automatically replaced by typing a new line with the

```
10 REM SIMBAS
20 DIM I(20),O(20),T0(20),T6(20)
51 PRINT
52 PRINT"TIME     A       B                 PLOT OF B"
54 K1=.2
55 K2=.1
56 T1=.2
57 T2=18
58 T3=1
80 N=2
86 O(1)=100
88 O(2)=0
98 T=INT(T2/T1+.5):T1=T2/T:T=INT(T3/T1+.5):T3=T*T1:T=0:T4=0
99 T8=1:GOTO1000
100 IF(T-T4+T1/10)<0 THEN 125
110 T4=T4+T3:T5=INT(T/T1+.5):T=T5*T1:GOSUB 2000
115 IF(T-T2+T1/10)<0 THEN 125
120 PRINT:PRINT"END":STOP
125 ON T8 GOTO 300,400,500,600
145 PRINT"** ERROR **":STOP
300 FOR T5=1 TO N
305 T0(T5)=T1*I(T5):T6(T5)=O(T5):O(T5)=O(T5)+T0(T5)/2
320 NEXT T5
330 T=T+T1/2:T8=2:GOTO1000
400 FOR T5=1 TO N
410 T7=T1*I(T5):T0(T5)=T0(T5)+2*T7:O(T5)=T6(T5)+T7/2
420 NEXT T5
430 T8=3:GOTO1000
500 FOR T5=1 TO N
510 T7=T1*I(T5):T0(T5)=T0(T5)+2*T7:O(T5)=T6(T5)+T7
520 NEXT T5
530 T8=4:T=T+T1/2:GOTO1000
600 FOR T5=1 TO N
605 O(T5)=T6(T5)+(T0(T5)+T1*I(T5))/6
610 NEXT T5
620 T8=1:GOTO 1000
1000    I(1)=-K1*O(1)
1010    I(2)=K1*O(1)-K2*O(2)
1999 GOTO 100
2000    B=25+O(2)/3
2010 PRINT T;O(1);O(2);TAB(B);"*"
2045 RETURN
```

Figure 2-1

LISTING OF SIMBAS

same number.

The example is a series of chemical reactions where A produces B that is a transient intermediate which reacts further. The rate equations are:

$$dA/dt = -K_1 A \tag{2.1}$$

$$dB/dt = K_1 A - K_2 B \tag{2.2}$$

BASIC notation is used with a special way of showing variables and derivatives. Derivatives (inputs to the integration operation) are denoted I(1), I(2), I(3), etc. (The letter I stands for IN). Dependent variables (outputs from integration) are denoted O(1), O(2), O(3), etc. (The letter O stands for OUT) In the example, I(1) stands for dA/dt and O(1) for A. Similarly, I(2) represents dB/dt and O(2) represents B. Thus, Statement 1000 is the dA/dt equation:

$$1000 \text{ LET } I(1) = -K1 * O(1) \tag{2.3}$$

Statement 1010 is the dB/dt equation:

$$1010 \text{ LET } I(2) = K1 * O(1) - K2 * O(2) \tag{2.4}$$

A second-order differential equation must be integrated twice to find the variable. For example,

$$d^2x/dt^2 = -32.2 \tag{2.5}$$

would require the following:

$$1000 \text{ LET } I(1) = -32.2 \tag{2.6}$$

$$1010 \text{ LET } I(2) = O(1) \tag{2.7}$$

where $I(1) = d^2x/dt^2$

$I(2) = dx/dt$

$O(1) = I(2) = dx/dt$

$O(2) = x$

Returning to the SIMBAS example, on Line 54, K1 is set to 0.2. On Line 55, K2 is set to 0.1. The reactions being simulated have a chemical, A, added to water. The initial concentration of A is 100 millimolar and B is initially zero. This is specified in Lines 86 and 88.

Time control of the solution depends upon the particular set of equations and on the portion of the solution which is of interest. Specifying a smaller integration interval, T1 in line 56, would give greater accuracy but at an increased cost of computer time and a longer wait for the answers. Different integration intervals, run times, and print intervals can be tried until satisfactory solutions are obtained. In general, select a small integration interval when variables are changing rapidly.

Output is controlled by two portions of the program. The headings are printed early in the program in Line 52 because they are wanted only once. The results are printed by Line 2000. In this case, time T, A, and B are written out at each print interval. Also shown is a TAB feature for plotting B; this works differently in some versions of BASIC. Those who can program in BASIC can devise means for getting the terminal to make crude graphs of several variables. Typical output of this example using the graphics features of SIM4 is shown in Fig. 2-2.

When all of the example lines have been replaced with new equations, coefficients, and values, a solution is initiated by typing RUN and striking the RETURN key. To halt a solution before completion, depress CNTRL while typing C. This may be slow in acting. Lines can be modified and replaced before trying RUN again. There are easier ways for modifying lines in BASIC than replacement, but the methods given here are sufficient for beginners. To save a program consult your BASIC manual.

The independent variable, T, may be accessed merely by using T in an equation.

Sources of Error

Small errors arise from numerical approximation of continuous functions, and these may propagate as the solution is extended. Furthermore, it is possible to introduce phase errors with SIMBAS. The user should exercise care in the order in which equations are written because a value needed in one equation may not be calculated until a later equation. The computer will use its last value from the previous integration interval thus risking error from the value being slightly out of phase with its use.

14

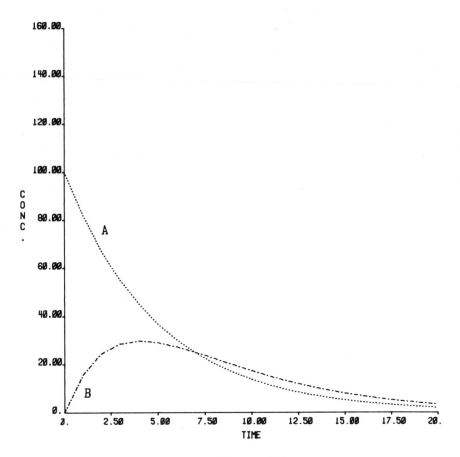

Figure 2-2
GRAPHS FOR SIMBAS EXAMPLE

OXYGEN DYNAMICS IN A FERMENTER

A simple exercise with SIM4 is modeling the response of a dissolved oxygen probe in a fermenter when the air supply is briefly interrupted. This simulation may be of some value as a classroom demonstration, but this author uses it to generate data for a student homework assignment. Only 10 to 20 seconds are required at the computer terminal to get a graph of dissolved oxygen versus time and a table of data. Thus a few minutes suffice for producing customized homework problems that are different for each student.

Simple mass balance considerations for the concentration of dissolved oxygen are:

$$dC/dt = K_l a \, (C^* - C) - r X \qquad (2.8)$$

where C = the concentration of dissolved oxygen
 t = time
 $K_l a$ = grouped transfer coefficient
 r = the uptake coefficient by the organisms
 C^* = oxygen concentration if saturated
 X = the concentration of organisms

When the air is turned off, there is no supply term, and the equation becomes:

$$dC/dt = - r X \qquad (2.9)$$

Data taken immediately after aeration ceases should be ignored because some air bubbles may be entrained in the medium to provide some oxygen. Quickly, the concentration of dissolved oxygen assumes a linear decline with a slope equal to $-r X$. If there has been an independent measurement of cell mass, the specific uptake coefficient can be calculated. Usually the goal is estimating $K_l a$, so the value of $r X$ is sufficient.

When aeration resumes, the supply term again becomes important. However, all terms in the equation except $K_l a$ can be found. The saturation value of dissolved oxygen can be estimated from published data or can be measured experimentally with broth free of organisms. The value of dC/dt is equal to the slope of the curve of C versus t. Students are asked to treat the data and to calculate $K_l a$. A typical graph is shown in Fig. 2-3. The relevant lines of the SIMBAS program are in Fig. 2-4, but the portions of SIMBAS that are very seldom of interest to the user are not shown. Note that IF statements are used to simulate the closing and

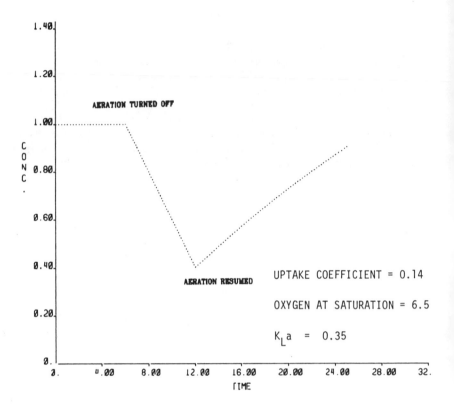

Figure 2-3
OXYGEN RESPONSE TO AN INTERUPTION IN AERATION

```
10 REM OXYGEN DYNAMICS
20 DIM I(20),O(20),TO(20),T6(20)
51 PRINT
52 PRINT"TIME    D.O.                PLOT OF D.O."
53 INPUT "R = "; K3
54 K1=0
55 K2=7.5
56 T1=.2
57 T2=25
58 T3=1
59 X=5
80 N=1
86 INPUT "I.C.= ";O(1)
1000 IF T>6 THEN K1=.2
1010 I(1)=K1*(K2-O(1))-K3*X
1999 GOTO 100
2000 B=10+10*O(1)
2010 PRINT T;O(1);TAB(B);"*"
2045 RETURN
```

Figure 2-4

RELEVANT PORTIONS OF SIMBAS FOR OXYGEN DYNAMICS

opening of the air valve at appropriate times.

VARIABLES THAT SHOULD NOT BECOME NEGATIVE

In the oxygen dynamics equations, the rate of oxygen uptake, (Equation 2.8), is based on no oxygen limitation. Of course, real concentrations cannot take on negative values, but computer simulations may go haywire. Misuse of equations often results in a variable reaching an impossible value. A more subtle error can result from too large an integration interval. The equations may be correct, but the variable may overshoot during the integration step and may stray into an impossible region. The system of equations may blow up when these values are introduced, and the simulation is meaningless.

A simple means of avoiding negative values is to use IF statements. For example,

60 IF X < 0 THEN X = 0.00001   (2.10)

Statement 60 checks the variable X and restores it to a small positive number should it go negative. This is better technique than setting to zero because there may be a division step that would halt the program because of a divide-by-zero error. Furthermore, microbial populations usually do not go to zero, and a small number represents reinoculation so that the population may recover at a later time. Usually, there will be no serious impairment of a simulation by preventing negative numbers because very low concentrations have little effect. Other factors should predominate or the main equations should drop out when a key concentration approaches zero.

TWO TANKS IN SERIES

For the next example, assume that two identical tanks are in series as shown in Fig. 2-5. The flow out of each is proportional to the level in that tank, and assume that the outlet valve positions are the same. For each, the mass balance is:

rate of change = flow in - flow out   (2.11)

and the individual equations are:

$dH1/dt = F - K\ H1$ (2.12)

$dH2/dt = K\ H1 - K\ H2$ (2.13)

where H1 and H2 are the respective liquid heights

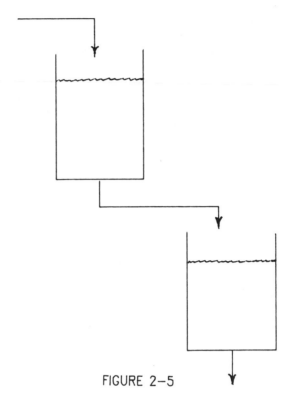

FIGURE 2-5

SKETCH OF TANKS IN SERIES

```
10 REM TWO TANKS IN SERIES
20 DIM I(20),O(20),T0(20),T6(20)
30 PRINT "TIME   HT. IN A   HT. IN B    PLOT FOR B"
51 PRINT
52 PRINT
54 K1=.3
55 K2=10
56 T1=.2
57 T2=25
58 T3=1
80 LET N=2
86 O(1)=0
88 O(2)=0
1000 I(1)=K2-K1*O(1)
1010 I(2)=K1*(O(1)-O(2))
1999 GOTO 100
2000 LET B=25+O(2)/3
2010 PRINT T;O(1);O(2);TAB(B);"*"
2045 RETURN
```

Figure 2-6

RELEVANT PORTIONS OF SIMBAS PROGRAM FOR TANKS IN SERIES

F = inlet flow to Tank 1
K = discharge coefficient

The relevant portions of the SIMBAS program for these equations are in Fig. 2-6, and solutions with the same values of F and K but with different initial conditions are in Figure 2-7. Note the contrast in initial slopes for filling the empty tanks.

Of course, several tanks could be in series and there could be complications such as non-uniform cross sections, non-linear relationships of discharge rate to head, interacting levels, and the like. In any case, the formulation of mass balances must include all of these effects. In real-world problems, concentrations of various substances in the fluid are usually of interest. It would be necessary to derive mass balance equations for each substance in each tank in addition to the equations for the fluid. Concentrations would depend on masses and on the volume of fluid. Although this can get complicated, the general approach is straight forward.

RESPIRATION WITH OXYGEN LIMITATION

Microbial respiration can be limited by either the availability of nutrient or availability of oxygen. A simulation of respiration in a fluid that receives oxygen by diffusion through an interface with air illustrates using IF statements to jump to a different equation depending on the oxygen concentration. There is also an IF statement to prevent the concentration of nutrient from becoming negative.

The analysis is much the same as used previously for oxygen dynamics, but the rate of removal of nutrient can be either a simple first-order decay or a zero-order reaction that depends on the oxygen concentration. Nutrient concentration, if expressed as biochemical oxygen demand (BOD), has one-to-one stoichiometry with oxygen.

Mass balance equations for dissolved oxygen and for nutrient are:

$$dC/dt = K_l a (C^* - C) + dL/dt \qquad (2.14)$$

$$dL/dt = -K1\ L \qquad (2.15)$$

or

$$dL/dt = -K2\ C \qquad (2.16)$$

where L = nutrient concentration
C = oxygen concentration
K1 = coefficient for nutrient decay

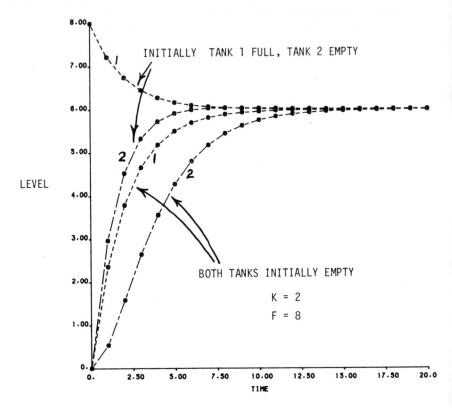

Figure 2-7

LEVELS IN TWO TANKS IN SERIES

K2 = coefficient when oxygen is limiting

The critical oxygen concentration is assumed to be 0.4, and this is slightly higher than typical values for real systems. Critical oxygen concentration is the point where oxygen availability becomes a factor. Because of the dramatic change in rate at the critical oxygen concentration, supply is able to match demand, and the oxygen concentration in an aerated fermentation tends to level off and not approach zero.

Relevant portions of a SIMBAS program for this situation are in Figure 2-8, and results with SIM4 are shown in Figure 2-9. The features of interest are the leveling off of oxygen concentration at about the critical concentration and its rise after the nutrient approaches exhaustion. The graph of nutrient concentration is somewhat uneven because of the switching action between equations.

SUMMARY

More examples will be provided later, but there should be sufficient understanding of simulation at this point for the reader to grasp the concepts for handling bioprocesses. Formulating differential equations is straightforward, and becomes quite easy with practice. The reader should actually work with the example programs and attempt some modifications and extensions to slightly more advanced problems.

```
10 REM SIMBAS FOR OXYGEN TRANSFER
20 DIM I(50),O(50),TO(50),T6(50)
40 PRINT
42 PRINT "TIME   ORGANICS   OXYGEN              PLOT OF OXYGEN"
50 C=7! :REM SATURATION VALUE OF OXYGEN
52 K1=.2 :REM COEFF. FOR DECAY OF ORGANICS
54 K2=.05:REM COEFF. WHEN LIMITED BY OXYGEN
56 K3=.12:REM KLA
57 T2=18
58 T3=1
59 T1=.2
80 N=2
86 O(1)=20!
88 O(2)=4!
1000 I(2)=K3*(C - O(2)) + I(1)
1005 IF O(1)<0 THEN O(1)=1E-03
1010 IF O(2)<.4 THEN GOTO 1050
1020 I(1)=-K1*O(1)
1030 GOTO 100
1050 I(1)=-K2*O(2)
1999 GOTO 100
2000 B=25 + 5*O(2)
2010 PRINT T;O(1);O(2);TAB(B);"*"
2045 RETURN
```

Figure 2-8
SIMBAS FOR FERMENTATION WITH OXYGEN LIMITATION

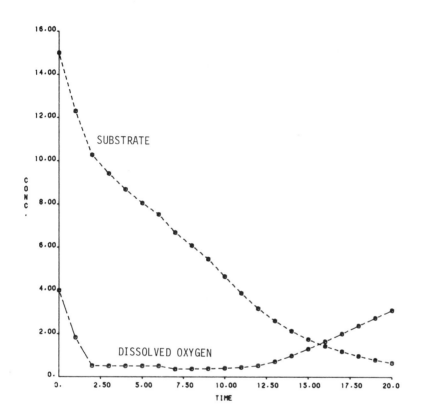

Figure 2-9

RESPIRATION WITH OXYGEN LIMITATION

SUGGESTED PRACTICE PROBLEMS

1. Write the SIMBAS equations for:

$$A \underset{3}{\overset{1}{\rightleftarrows}} B \xrightarrow{2} C$$

You will find it easy to modify the SIMBAS example to handle this problem. Determine the time behavior of each material with the following sets of rate coefficients:

a. K1 = 0.5, K2 = 0.8, K3 = 0.01

b. K1 = 0.5, K2 = 0.1, K3 = 0.8

c. your choice.

2. Write a SIMBAS equation for level in a cylindrical tank where the rate of discharge from the bottom is proportional to the square root of the level. Use your own choice of coefficients.

3. Write a SIMBAS equation for concentration of microorganisms in a vessel assuming no input of organisms and a constant specific growth rate coefficient, μ. Do not forget to provide an initial condition and start with a small growth rate.

Chapter 3

BATCH PROCESSES

For many years, the usual way of operating fermentations was to sterilize the medium, cool, inoculate, add air continuously, and add antifoam on demand. This was considered a batch operation even though air and antifoam were fed. A major change in thinking occurred because of improvements in the penicillin fermentation. An expensive sugar, lactose, was the carbohydrate source, but the effectiveness of lactose was found to result from its slow rate of hydrolysis to glucose and galactose by the penicillium culture. There was a good match to the growth rate of the organisms so that the concentration of monosaccharides stayed low. If the monosaccharide level is not low, the growth phase is prolonged and penicillin synthesis is impaired. Feeding of glucose to control the concentration at a desired level was more effective than relying on the enzymatic hydrolysis of lactose, and industrial fermentations experienced a 3 to 4-fold improvement in yield. Greatly reduced sales of lactose caused its price to fall to about that of glucose, but a major market was lost. With this dramatic demonstration of the value of feeding, the fed batch fermentation gained much attention and has become the most popular industrial method.

A computer game is useful in teaching some characteristics of the older batch fermentation system (Bungay, 1971). It is patterned after a typical pilot plant assignment with sequential runs in small fermenters where the recipe for the medium and the conditions must be specified. As part of an academic course, the game provides an excellent frame of reference in that the players gain a good grasp of a non-growth related fermentation, of the relation of antifoam usage to aeration, and of the costs. They also learn about common nutrients and begin to appreciate the complexities of interactions in recipes.

FERMT, A FERMENTATION PROCESS DEVELOPMENT GAME

The objective of this game is to optimize profit. A player is allowed to manipulate seven variables, but has only ten small vessels (fermenters) with which to experiment. Expected yield and profit are computed for a full-scale production batch. The player thus submits ten different recipe formulations and gets back some data and profit figures. The problem is to use this information to plan the next set of ten recipes. After eight pilot trials, the process will go into full production. The profit to expect is known to the instructor, and students are graded on the logic of their approaches plus a bonus if their profit is superior to the class average. Some typical results are shown in Fig. 3-1.

Each of the seven variables can be beneficial at a certain level and harmful in excess. The process is two-stage in that growth and product formation occur separately. Some things which are good for growth are bad for product formation, and several of the variables interact in producing effects. The computer program for the model is not shown to the player.

Variables to be manipulated: (All ingredients are per cent by weight.)

Inoculum. This is expressed as per cent of seed culture added to the total culture volume. No growth is possible without seed, but seed culture is expensive. The problem is to get off to a quick start by adding sufficient seed but to avoid the cost of using too much.

Sugar. Adding more sugar results in more growth but this prolongs the growth phase. Since product formation commences only after the growth phase is over, there must be a compromise between getting abundant growth and having sufficient time for this growth to do something. Scheduling restraints for the plant define a fixed total time for a run.

Oil. Oil is a nutrient which can substitute for sugar except that excess oil is harmful to both growth and to product formation. Oil also serves as an antifoam agent. High air flows promote foaming and greatly increase oil consumption.

Soymeal. This is a complex nitrogenous material. It contains some vitamins, other undefined essential ingredients, and toxic ingredients. Some soymeal is beneficial; too much is detrimental.

Distillers solubles. This is also a complex nitrogenous material. It contains some of the same factors as soymeal and

FERMENTATION GAME

INOCULUM	SUGAR	OIL	SOYMEAL	DISTSOL	AIR	VITAMINS
1.0000	4.0000	1.0000	0.1500	0.1500	1.0000	0.0100

PRODUCT FORMATION BEGAN AT 56.00 HOURS
TOTAL GROWTH= 4.802   TOTAL OIL = 3.880   YIELD = 1207.092
PROFIT = $ 143720.52

INOCULUM	SUGAR	OIL	SOYMEAL	DISTSOL	AIR	VITAMINS
2.0000	4.0000	1.0000	0.1500	0.1500	1.0000	0.0100

PRODUCT FORMATION BEGAN AT 49.00 HOURS
TOTAL GROWTH= 4.711   TOTAL OIL = 3.880   YIELD = 1391.375
PROFIT = $ 154302.94

INOCULUM	SUGAR	OIL	SOYMEAL	DISTSOL	AIR	VITAMINS
5.0000	9.0000	1.0000	0.2500	0.2000	0.9000	0.0005

PRODUCT FORMATION BEGAN AT 52.00 HOURS
TOTAL GROWTH= 8.386   TOTAL OIL = 2.890   YIELD = 867.259
PROFIT = $ -12403.27

INOCULUM	SUGAR	OIL	SOYMEAL	DISTSOL	AIR	VITAMINS
5.0000	8.0000	1.0000	0.2000	0.2000	0.9000	0.0050

PRODUCT FORMATION BEGAN AT 50.00 HOURS
TOTAL GROWTH= 7.606   TOTAL OIL = 2.890   YIELD = 930.109
PROFIT = $ -448.19

INOCULUM	SUGAR	OIL	SOYMEAL	DISTSOL	AIR	VITAMINS
10.0000	5.0000	1.0000	0.2000	0.2000	0.9000	0.0005

PRODUCT FORMATION BEGAN AT 36.00 HOURS
TOTAL GROWTH= 5.248   TOTAL OIL = 2.890   YIELD = 1289.020
PROFIT = $ -49100.67

INOCULUM	SUGAR	OIL	SOYMEAL	DISTSOL	AIR	VITAMINS
8.0000	6.0000	0.5000	0.0500	0.0500	0.7000	0.0100

PRODUCT FORMATION BEGAN AT 42.00 HOURS
TOTAL GROWTH= 5.335   TOTAL OIL = 1.191   YIELD = 2185.693
PROFIT = $ 162409.32

INOCULUM	SUGAR	OIL	SOYMEAL	DISTSOL	AIR	VITAMINS
9.0000	8.0000	0.5000	0.1900	0.1000	0.7000	0.0100

PRODUCT FORMATION BEGAN AT 47.00 HOURS
TOTAL GROWTH= 6.911   TOTAL OIL = 1.191   YIELD = 1802.017
PROFIT = $ 68117.89

INOCULUM	SUGAR	OIL	SOYMEAL	DISTSOL	AIR	VITAMINS
1.0000	5.0000	0.1000	0.1000	0.1000	1.0000	0.0200

PRODUCT FORMATION BEGAN AT 56.00 HOURS
TOTAL GROWTH= 4.805   TOTAL OIL = 2.980   YIELD = 1530.322
PROFIT = $ 203232.34

Figure 3-1
TYPICAL RESULTS FOR FERMT GAME

some other factors. The same comments apply as for soymeal.

Air. This aerobic process requires air during both the growth and production phases. Although air is cheap, high aeration is harmful if oil consumption for antifoam is excessive. Units are volumes of air per minute per volume of batch (VVM). More than about 2 VVM would probably be impossible for a real process. While there is no upper restraint on this hypothetical process, antifoam oil usage becomes astronomical.

Vitamins. This term is applied to all the vitamins and trace elements added to this fermentation. Small amounts are found in soymeal and distillers solubles, but some supplementation may be desirable. Excess vitamins reduce the rate of product formation.

Profit is calculated by subtracting the cost of the ingredients and operating costs from the value of the product. Negative concentrations are illegal, and the computer just prints that the run was contaminated if specifications are out of line. Common sense is of some help to the player in that inoculum cannot be zero and vitamins are used in much smaller amounts than is sugar.

HINTS:

1. Use more sugar than nitrogenous compounds.

2. Too much of anything is bad.

HAND IN:

1. Some representative results.
2. A very brief explanation of your reasoning.

A listing of the FERMT game in BASIC is in Appendix 2 and a FORTRAN listing is also provided. The coefficients are usually changed each year so that old results are of little help. There are no graphics features, so the game is played on personal computers or on the main campus computer. One convenience on both small and large computers can be visual editing where all the recipes are displayed and the player need only move the cursor of the terminal and retype whatever specifications need to be changed.

A very clever instruction sheet for the FERMT game was written by M.R.J. Morgan of the University of Lancaster. His group also translated this game from FORTRAN to BASIC with a program equivalent the listing in Appendix 1. A witty introduction to a game helps motivate the students, and British writers seem especially good at this.

JERMFERM, A FERMENTATION PROCESS GAME

The general concept of FERMT has been used for a more elaborate game (Mateles, 1978). This game, JERMFERM, allows the player to specify: concentrations of glucose, soybean meal, phosphate, and salts; agitation and aeration rates; fermenter operating volume; inoculum size; and duration of the fermentation. JERMFERM has somewhat different teaching objectives than FERMT by seeking greater insight into fermentation details and for training in pH and buffering effects. Whereas FERMT tries to show the role of the scientist or engineer in the pilot plant, JERMFERM is a tool for studying individual fermentations. JERMFERM has no time restraints and lets the player examine the results from each fermenter before specifying the next run. It would be unwise to compare the games because each is better for its particular purpose.

A sequential search technique has been reported for optimizing JERMFERM (Saguy, 1982). This way to "beat the game" is an interesting, systematic exercise, but the tedious manipulation of the many variables probably is more educational for someone trying to gain insight into a fermentation process.

An RPI graduate student, Steven Fraleigh, converted JERMFERM to interactive computer graphics. Another student, Paul Raymond, wrote an instruction file for playing the game. A typical run is shown in Figure 3-2. The light pen is touched to a variable, and the computer accepts a number for its value. Note that the graphs, the menu, the inputs, and the cost information are displayed simultaneously. Previous runs can be recalled for comparison. A new run is initiated by touching the light pen to RESET and then changing one or more input specifications. The box in the lower right-hand corner keeps track of which run gave the best daily profit. Interactive graphical presentation seems to be highly effective for speeding up this game while improving communication to the player.

SYNCHRONOUS CULTURE

Properties of very large numbers of cells can be treated as continuous functions. There is a broad spectrum of cell ages, and cell growth occurs as small increments that meld into smooth curves for anything that is measured. In other words, a single event for a cell is such a tiny blip for the many, many cells present that everything averages out to

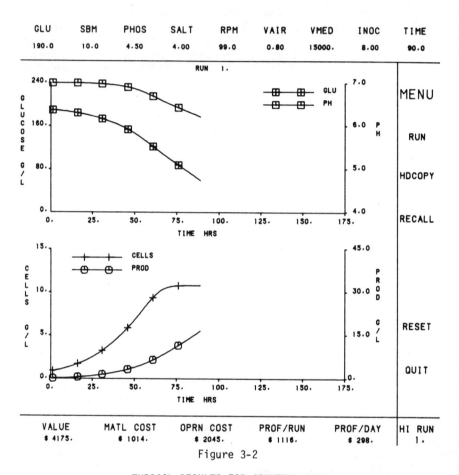

Figure 3-2

TYPICAL RESULTS FOR JERMFERM GAME

continuous functions. There are exceptions. When individual cells are modeled, difference equations may be needed instead of differential equations. Very small populations may sometimes require stepwise treatment, but this is very uncommon. As the cells replicate, total numbers become so great that differential equations are quite suitable. Synchronous culture is an important case where care must be used in employing differential equations. Cells can all be induced to divide at the same time if illumination or temperature or some other factor can impair or halt a step in division. The cells can be triggered to proceed together from that point with overall numbers that are stepwise with time. This is termed a synchronous culture; the steps are seldom distinct for more than a few generations unless the triggering event continues to be applied periodically.

A rather simple-minded approach to modeling synchronous culture is to assume that substrate uptake will be proportional to the number of organisms. The equation for rate of change of substrate is:

$$dS/dt = - K * X \qquad (3.1)$$

where X = concentration of organisms
S = concentration of substrate
K = uptake coefficient for substrate

The concentration of organisms is assumed to double at regular time intervals. Portions of a SIMBAS program for this simulation are in Figure 3-3, and typical results from a SIM4 run are shown in Fig. 3-4. Note the stepwise increases in X while S is continuous but made up of linear segments. Although X doubles instantaneously, the graphs show a slope for each step because of the way that points are connected by lines. Such artifacts of the simulation technique can be quite misleading.

MATHEMATICS OF BATCH CULTURE

Mass balances for common, unsynchronized batch culture give:

$$dX/dt = \mu X - K_d X \qquad (3.2)$$

$$dS/dt = -\mu X / Y \qquad (3.3)$$

$$\mu = f(S) \qquad (3.4)$$

```
10 REM SYNCHRONOUS CULTURE, X INCREASES IN STEPS
20 DIM I(20),O(20),T0(20),T6(20)
40 X=1
51 PRINT
52 PRINT "TIME        X            S"
54 K1=.4
56 T1=.2
57 T2=25
58 T3=1
59 Y=5
80 N=1
86 O(1)=100
1000 I(1)=-K1*X
1999 GOTO 100
2000 IF T<Y THEN GOTO 2100
2010 X=2*X
2020 Y=Y+5
2100 PRINT T,X,O(1)
2900 RETURN
```

Figure 3-3
PORTION OF SIMBAS PROGRAM FOR SYNCHRONOUS CULTURE

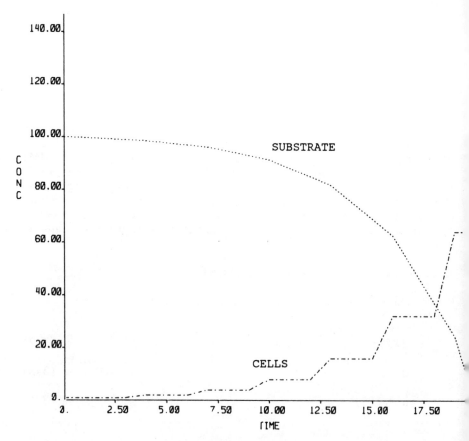

Figure 3-4
SIMULATION OF SYNCHRONOUS CULTURE

where X = concentration of organisms
       S = concentration of substrate present in lowest proportion
       μ = specific growth rate coefficient
       K_d = death rate coefficient
       Y = yield coefficient, mass of cell per mass of S

Various functional relationships between μ and S are used, but the Monod equation is by far the most popular:

$$\mu = \hat{\mu} \frac{S}{K_s + S} \tag{3.5}$$

where $\hat{\mu}$ = maximum growth rate coefficient
       K_s = Monod coefficient

The death rate coefficient is usually relatively small unless inhibitory substances accumulate, so equation 3.2 shows an exponential rise until S becomes depleted to reduce μ. This explains the usual growth curve (Fig. 3-5) with its lag phase, logarithmic phase, resting phase, and declining phase as the effect of Kd takes over.

STRUCTURED MODELS

Meaningful detail can be added to culture models in several ways. Cells can be compartmentalized according to biochemical functions, and the components can interact. For example, there can be a group of equations for carbohydrate metabolism, a group for protein synthesis, another for nucleic acid synthesis, etc. This permits a much more intricate description of cellular activities but at the expense of having so many rate constants that assigning values to them may end up as guesswork. For cells with distinct life cycles, a structured model may have compartments corresponding to each stage in the cycle. In addition, each compartment may be subdivided into the biochemical functions mentioned above. Such complicated models have had limited practical use but have great value for directing research toward areas where information is lacking.

Ramkrishna (1979) provides elegant and thorough analysis and discussion of structured models with emphasis on statistical aspects. He points out that experimental verification is becoming possible through such studies as

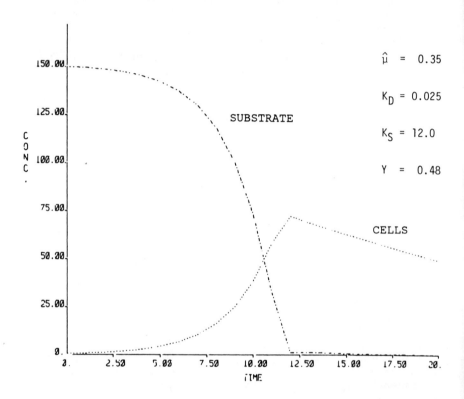

Figure 3-5
SIMULATION OF BATCH FERMENTATION

those by Bailey, et al (1977) in which flow cytometry of individual cells measures the concentrations of various biochemicals.

FED BATCH SYSTEMS

Development of a fed batch system is difficult because several variables must be manipulated. There are interactions between timing of feed, concentration of feed, and the physiology of the culture. Some cases may simply require keeping a nutrient at a fixed concentration, and the equations for organism and substrate concentrations are the same as used previously. Although more sophisticated control schemes will be discussed later, on-off control will illustrate this type of fed batch reasonably well. The relevant portion of a SIMBAS program is shown in Fig. 3-6. The concept is to prevent S from falling below a specified concentration. Note the use of IF statements for controlling sugar feed. Results of this simulation are shown in Fig. 3-7. The growth rate changes as the substrate level falls.

Bosnjak, et al. (1979) have reported on growth kinetics and antibiotic synthesis in repeated fed-batch culture of Streptomyces rimosus. Various feed rates could be tested, and their model agreed with experimental data. Aiba (1979) used on-line computer control for fed-batch culture of Saccharomyces cerevisiae. The feed rate was based on respiration coefficient. Peringer and Blanchard (1979) modeled fed-batch production of Bakers' yeast and devised an optimal control system. Again, respiratory coefficient was the basis for selecting the feed rate. An interesting feature of this study was consideration of catabolite repression, the inhibition of growth by excess sugar concentration. Weigand, et al. (1979) used an elegant analysis of fed-batch fermentation in which the yield coefficient could be allowed to vary and either Monod growth kinetics or a function for inhibition of growth could be employed.

SEMI-CONTINUOUS CULTURE

Inoculation of a fermenter could be avoided by adding fresh nutrient medium to a portion of the broth from a previous run. In other words, most of a run could be harvested while leaving some in the fermenter to seed the next run. There are only a few practical examples of this technique because such prolonged operations provide opportunities for contamination, and high-yielding mutant cultures would probably revert over such long times to less productive forms. One fermentation that can employ this "back-seeding" technique is for fuel-grade ethanol because quality control is relatively unimportant, and yeast are hardy

```
10 REM FEDBATCH SYSTEM, ADD SUGAR WHEN LOW
20 DIM I(20),O(20),TO(20),T6(20)
51 PRINT
53 PRINT"TIME    X     S              PLOT OF S"
54 K2=1
55 K3=5:F=1:Y=.55
56 T1=.02
57 T2=18
58 T3=1!
80 N=2
86 O(1)=1
88 O(2)=4
1000 I(1)=K1*O(1)
1010 I(2)=F-I(1)/Y
1020 K1=K2*O(2)/(K3+O(2))
1999 GOTO 100
2000 B=25+O(2)
2005 F=0
2007 IF O(2)<2 THEN F=40
2010 PRINT T;O(1);O(2);TAB(B);"*"
2045 RETURN
```

Figure 3-6

RELEVANT PORTIONS OF SIMBAS FOR FED BATCH FERMENTER

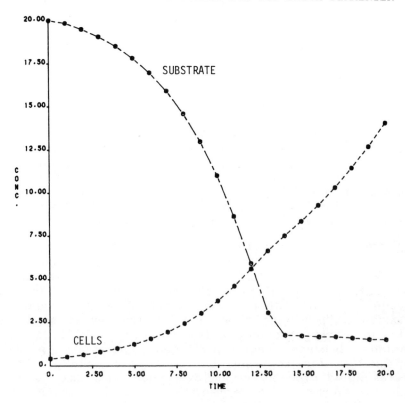

Figure 3-7

BATCH CULTURE WITH SUGAR FEED

```
2 REM BACK SEED SIMULATION
20 DIM I(20),O(20),TO(20),T6(20)
30 REM D IS DEATH COEFF., P+1 IS HARVEST PERIOD
40 D=.01
41 P=9
50 PRINT "TIME    X      S         PLOT OF X"
51 PRINT
56 T1=.2
57 T2=60
58 T3=1
80 LET N=2
86 O(1)=1
88 O(2)=50
1000 I(1)=O(1)*(M-D)
1010 I(2)=-2*M*O(1)
1020 M=.5*O(2)/(4+O(2))
1999 GOTO 100
2000 B=30+O(1)
2010 PRINT T;O(1);O(2);TAB(B);"*"
2020 IF T>P THEN GOTO 2050
2045 RETURN
2050 P=P+10
2060 O(1)=O(1)/10
2075 T6(1)=O(1)
2080 O(2)=41
2095 T6(2)=O(2)
2999 RETURN
```

Figure 3-8
PORTION OF SIMBAS PROGRAM FOR BACK SEEDING

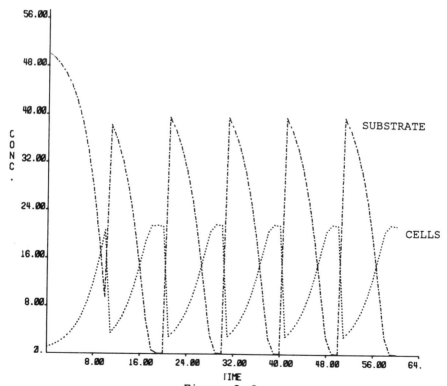

Figure 3-9
SIMULATION OF BACK-SEED FERMENTATION

organisms that thrive at the low pH and alcohol concentrations that discourage contaminants.

The relevant portions of a SIMBAS program for this mode of operation are in Fig. 3-8. Note that the stored variable, T6(1), must also be reset when the value of the cell concentration is reset. Results of a simulation of semi-continuous fermentation are shown in Fig. 3-9. The periodic nature is obvious. It would not be difficult to add complications such as changes in the Monod coefficients from run to run, but such exercises have questionable value unless related to actual processes.

Combining fed batch and semi-continuous techniques is an easy and logical extension of the simulations already presented. This is a suggested exercise for the reader.

REFERENCES

Aiba, (1979) S., "Review of Process Control and Optimization in Fermentation", Biotechnol. Bioengr. Symp. 9, 269-281

Bailey, J.E., J. Fazel-Madjlessi, D.N. McQuitty, D. Lee, and D. Oro, (1977) "Characterization of Bacterial Growth Using Flow Cytometry", Science 198: 1175

Bosnjak, M., V. Topolovec, and V. Johanides, (1979) "Growth Kinetics and Antibiotic Synthesis during the Repeated Fed-batch Culture of Streptomyces", Biotechnol. Bioengr. Symp.9, 155-165

Bungay, H.R., (1971) "FERMT, A Fermentation Game Based On Process Development", Process Biochemistry 6: 38

Mateles, R.I., (1978) "JERMFERM, A Fermentation Process Development Game", Biotechnol. Bioeng. 20: 2011-2014

Peringer, P., and H.T. Blanchere, (1979) "Modeling and Optimal Control of Bakers' Yeast Production in Repeated Fed-batch Culture", Biotechnol. Bioengr. Symp. 9, 205-213

Ramkrishna, (1979) D., "Statistical Models of Cell Populations", Adv. Biochem. Engr. 11: 1-47

Saguy, I., (1982) "Utilization of the 'Complex Method' to Optimize a Fermentation Process", Biotechnol. Bioengr. 24: 1519-1525

Weigand, W.A., H.C. Lim, C.C. Creagan, and R. D. Mohler, (1979) "Optimization of a Repeated Fed-batch Reactor for Maximum Cell Productivity", Biotechnol Bioengr. Symp. 9, 335-348

Chapter 4

CONTINUOUS CULTURE

Continuous culture has been the goal of bioengineers for several decades because batch culture has inherent down time for cleaning and sterilization and long lags before the organisms enter a brief period of high productivity. With continuous culture, the nutrition and the product mix can be advantageously manipulated as functions of dilution rate. A serious problem, however, is genetic instability of the culture itself. Finely tuned mutants that produce high titers of product may revert to less productive strains. The main successes with continuous fermentation have been with rugged strains that are producing either cell mass or a simple enzyme or metabolite. When a single stage is used, it is difficult to optimize cell growth, product production, and good utilization of substrates. Molds used for antibiotic fermentations are particularly messy and form voluminous coatings on the shaft, coils, and any protuberances in the fermenter. Although this may limit the length of a run, the more important problem is culture rundown.

The usual explanations for limited use of continuous culture in industry are: culture instability, difficulty of maintaining asepsis, insufficient knowledge of microbial behavior, and reluctance to convert existing factories. Another more subtle reason is the cost of each research station. Rapid progress in research and development requires multiple vessels for screening many variables, but there are usually only one or two continuous fermenters because the cost of pumps, reservoirs, sterilizers, and controls is relatively high. Furthermore, much labor is needed to start a run, to keep the instruments working, and to monitor results.

Continuous culture is usually superior to batch culture for research. Batch culture suffers from changing concentrations of products and reactants, varying pH and redox potential, and a complicated mix of growing, dying, and dead cells. Data from continuous cultures are much easier to interpret because steady states are achieved or there are repeatable excursions from steady state.

Conventional means for continuous culturing are the chemostat in which nutrient is fed to a reactor at constant rate, and the turbidostat which employs feedback control of pumping rate to maintain a fixed turbidity of the culture. Another alternative with feed back control of a nutrient or of product concentration has been termed controlled, concentration-coupled, continuous culture or "C5". Proportional control of the pumping rate is desirable because continuous cultures can have oscillatory responses induced by turning the feed pump on or off. Our group has found that "C5" continuous cultures have a powerful selective advantage for rapidly growing organisms. After several days of operation, strains may be selected that double in less than 10 minutes (Bungay, et al 1981).

MATHEMATICAL ANALYSIS

The concept of a limiting nutrient is essential to the theory of continuous culture. The ingredient in short supply relative to the other ingredients will be exhausted first and will thus limit cellular growth or product synthesis. The other ingredients play various roles such as exhibiting toxicity or promoting cellular activities, but there will not be an acute shortage as in the case of the limiting nutrient. Some typical references about continuous culture are Malek and Fencl (1966) and Bailey and Ollis (1977).

The analysis of continuous cultivation of microorganisms starts with the mass balance equation:

rate of change = input - output + reaction   (4.1)

In the special case of the device termed a chemostat, there is a selected, constant rate for feeding sterile medium to organisms in a vessel of constant volume. This means, of course, that the feed rate equals the overflow rate. The mass balance for the organisms is:

$$dx/dt \, V = F \, x_i - F \, x + \mu \, x \, V \qquad (4.2)$$

where x = organism concentration in g/L at time t
       V = the constant volume of the vessel
       F = feed rate
       μ = specific growth rate coefficient
       x_i = organism concentration in the feed

In most continuous culture systems, the feed stream is sterile, so the organism input term = 0. The only notable exceptions are waste treatment systems. With excellent mixing, x in the vessel is assumed the same as x in the overflow. A new term D, the dilution rate, is equal to F/V, thus dividing Equation 4.2 by V and dropping the zero term gives

$$dx/dt = \mu x - D x \qquad (4.3)$$

A mass balance for the nutrient in lowest proportions for growth (the growth-limiting nutrient) gives:

$$dS/dt\, V = F S_o - F S - \mu x V/Y - M x V \qquad (4.4)$$

where S   = the concentration of limiting nutrient, g/l
       S_o = the concentration in the feed stream
       Y   = the yield coefficient, g of cells/g of limiting nutrient
       M   = the maintenance coefficient to keep cells alive

Dividing by V, we get:

$$dS/dt = D S_o - D S - \mu x/Y - M x \qquad (4.5)$$

A relationship is needed for µ and S, and this is the Monod equation:

$$\mu = \hat{\mu} \frac{S}{K_s + S} \qquad (4.6)$$

where $\hat{\mu}$ = the maximum specific growth rate with excess S
       K_s = a constant

One advantage of the chemostat type of operation is that there is a tendency to stable steady state. At steady state, there is no change thus the derivatives in the differential equations disappear to give:

$$\mu = D \qquad (4.7)$$

and

$$D S_o - D S = \mu x/Y + M x \qquad (4.8)$$

Substituting D for µ in Equation 4.6 and solving for S gives:

$$S = \frac{D\,K_s}{\hat{\mu} - D} \tag{4.9}$$

Solving Equation 4.8 for x after substituting D for μ gives

$$x = D\,Y\,\frac{S_o - S}{D + M\,Y} \tag{4.10}$$

This is an old, familiar analysis that applies to any continuous culture that meets the assumptions of perfect mixing and constant volume. The equations are fundamental except for the Monod equation which has no time dependency and should be applied with caution to transient states where there may be a time lag as μ responds to changing S.

Mixing has been shown to be critical at low dilution rates because uptake of substrate is extremely rapid for cells in a starved condition. Quite vigorous agitation is required in small vessels to insure homogeneous distribution of the feed; such intense agitation is probably impractical in large vessels, but it might be possible to distribute the feed from many fine openings throughout the tank.

MONOD, A Single-Stage Chemostat Game

The MONOD teaching game was conceived late one afternoon, programmed in one hour, and tested with a student before the end of the day. Despite this hasty invention, it has been tried and true as a teaching tool for over twelve years. It is now more elegant with computer graphics features, but rough page plots for the terminal or a printer can still be quite effective. A BASIC version is in Appendix 3.

The theory of continuous culture becomes very clear when illustrated by computer graphics. Lectures about the mathematics of continuous culture are easy to understand, but it is difficult to visualize the effects of the various coefficients without graphical presentations. MONOD ties together the topic of continuous culture very nicely so that the theory achieves real meaning. The mathematical basis is the same as presented earlier, and coefficients specified by the player are substituted into:

$$S = \frac{D\,K_s}{\hat{\mu} - D} \quad \text{repeat of (4.9)}$$

and

$$x = \frac{D\,Y\,(S_o - S)}{D + M\,Y} \quad \text{repeat of Equation (4.10)}$$

Note that the values needed are $\hat{\mu}$, K_s, Y, S_o, and M. The computer increments D until it is greater than $\hat{\mu}$ at which point organisms would wash out. Starting with D equal to zero would cause division by zero in Equation 4.10 if M is specified as zero, thus a small value is added to D and later subtracted before displaying the results. This also prevents a divide-by-zero error in Equation 4.9 when D approaches μ because the increment is chosen unequal to the small value carried along with D.

Reasonable values for the parameters when using the scales on the graph for the R.P.I. program are:

$\hat{\mu}$	0.1 to 3.0 hr-l
Y	0.2 to 0.7 g/g
S_o	1 to 100 g/L
K_s	0.2 to 10.0
M	0.001 to 0.2

Other ranges might be better for other process such as algal systems or biological waste treatment where growth rates are slow and nutrients are very dilute.

The following instructions are given to the players:

1. Study the effects of increasing and decreasing $\hat{\mu}$, Y, So, and Ks while keeping M very small. Explain your observations. Use M = 0.001.

2. Select a set of the other parameters that produces curves that fit the plotting area well and experiment with different values of M. Explain the results.

3. Hold M at a relatively high value and then at a low value while exploring ranges of other parameters. Explain your results.

Some typical results with the MONOD game are shown in Fig. 4-1. Several plots may be overlayed for easy comparisons, and a cluttered screen can be erased. The most important concepts from the game are:
1. x goes to zero and S reaches S_o as D approaches $\hat{\mu}$,
2. S is not a function of So when D is less than $\hat{\mu}$,
3. the maintenance coefficient is very important but only at low dilution rates.

44

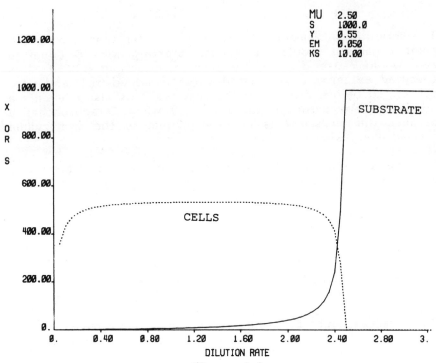

Figure 4-1
TYPICAL OUTPUT FOR MONOD GAME

```
D                PLOT OF X AND S VERSUS D

0    SX
.1   S                              X
.2   S                              X
.3   S                              X
.4   S                              X
.5   S                              X
.6   S                              X
.7   S                              X
.8   S                              X
.9    S                             X
1     S                             X
1.1   S                             X
1.2   S                             X
1.3    S                            X
1.4    S                            X
1.5     S                           X
1.6     S                           X
1.7      S                          X
1.8       S                        X
1.9        S                       X
2           S                    X
2.1          S                 X
2.2            S         X
2.3              X       S
2.4        X                                     S
2.5 X
2.6 X
```

Figure 4-2
PRINTER PLOTS FOR MONOD GAME

The MONOD game is very well suited to personal computers with color graphics. An acceptable plot for terminals without graphics features can be obtained by spacing characters on the screen as shown in Figure 4-2. Some simple logic must be incorporated into the program to decide the character to be printed first because the graphs cross.

For extra credit on this assignment, students can explore modifications in the basic assumptions of the simulation. For example, the yield coefficient should not be constant. The average age of organisms in a chemostat depends on dilution rate. Average age is high at low dilution rates, but young cells predominate at high dilution rates where there is a high probability of washout before reaching advanced age. It is very easy to postulate a relationship between Y and D and to incorporate it into the computer program. In like manner, the Monod equation coefficients can be made functions of D. This assignment can have considerable value when the student goes to the literature to find support for the new relationships because there is an appreciation of the importance of the coefficients and of the paucity of available information.

RECYCLE

Separation and recycle of cells results in much longer residence times for the cells than for the fluid and permits relatively high cell concentrations. In waste treatment, the dilute feed leads to slow growth rates. More rapid processing is attained by achieving higher populations through cell recycle although a higher percentage of the cells may be dead if all are in a starved state. High rates of production are also important in industrial fermentations with cell recycle, and there is the added advantage of reusing cells instead of diverting expensive substrate to replace those that leave in the harvested broth.

In activated sludge, organisms not associated with flocs are not collected and tend to leave the system as recycle increases the proportions of flocculating types. Recycle of collectible algae to outdoor ponds has profound influence on the population, but seasonal changes can develop small algae despite retention of large algae, thus recycle fails as the few large algae die or escape collection.

Recycle of fermentation fluids has quite different objectives than those for cell recycle that aims for population control or greater productivity. Spent broths have left over nutrients so recycle can save on costs of nutrients and make up water while greatly reducing the volumes sent to waste treatment. Of course, total recycle is bad because undesirable materials accumulate and can poison the fermentation. This build up determines the amount of recycle, but there may be purification steps to remove toxic substances. For example, acetic acid is a byproduct of alcohol fermentation but its low volatility means that much is in the stillage from alcohol recovery. Its removal by a physical or chemical step would prevent accumulation in the fermenter by recycle. Alternatively, it could be metabolized by a special strain or mixed culture.

Mass balances for a continuous culture vessel with recycle are:

$$F_y = (1 + \omega) F \tag{4.11}$$

$$F = F_e + F_{ex} \tag{4.12}$$

$$F_y = F_e \frac{Xe}{X} + \frac{Xx}{X} (F_{ex} + \omega F) \tag{4.13}$$

where F_y = flow rate leaving vessel
F = rate of fresh feed
ω = recycle ratio
F_e = rate of liquid stream
F_{ex} = rate of slurry take off
X = cell concentration in reactor
Xe = cell concentration in liquid
Xx = cell concentration in slurry

Assuming perfect mixing:

$$V \, dX/dt = Xx \, \omega \, F - F_y X + V \mu X \tag{4.14}$$

Combining equations and setting the derivative to 0 for steady state:

$$\mu = (1 + \omega) D - \omega D \, Xx/X \tag{4.15}$$

This leads to:

$$\mu = D + D\omega \left(1 - \frac{1 + \omega - Fe/F - Xe/X}{1 + \omega - Fe/F}\right) \tag{4.16}$$

Without recycle, washout occurs when D is greater than the maximum specific growth rate. With recycle, D can greatly exceed the maximum specific growth rate.

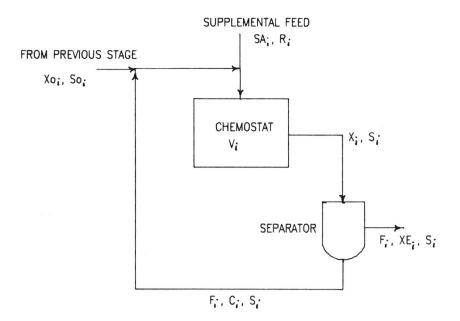

FIGURE 4-3
ELEMENT OF A MULTI-STAGE SYSTEM

MULTI-STAGE CONTINUOUS CULTURE

It is easy to postulate advantages for multistage continuous culture but very difficult to conduct all of the research and development of the many parameters that should be optimized. Each stage could have its feed streams, control of pH and other conditions, and recycle of cells or fluids from other steps in the process. Not only are there many parameters to study for each stage, but changes in one stage can markedly affect other stages. It can be quite troublesome to get representative conditions and cultures in a given stage to begin research because of the complicated interactions with other stages. Time delays in lines and in separators for recycle plus complexities from non-ideal flow regimes cause theoretical analysis to be faulty. An optimized multistage continuous fermentation system with recycle and control must be one of the most difficult engineering feats. To complicate matters even more, the dynamics of microbial responses to upsets are poorly understood.

For his student project in one of our courses in biochemical engineering, H. Y. L. Lam became intrigued by multi-stage continuous culture and went over the usual time for a term project by a factor of about 20. The result was an elegant simulation game that is useful for both research and teaching (Lam and Bungay, 1983). Other students have tried to invent new games, but few have achieved any success.

DESCRIPTION OF THE PROGRAM

Fundamental mass balances for organisms and for substrate in a perfectly mixed reactor are formulated as differential equations and set to zero to solve for steady state concentrations. Amounts entering can be exit streams from the previous stage, a recycle stream, or a new feed stream of nutrient medium. The amount leaving is simply the concentration times the flow rate of the effluent. The substrate equations account for uptake by organisms, amd cell mass is created by growth. Specific growth rate coefficient for the organisms has been taken from the Monod equation which usually works very well for most microbial systems at steady state.

When the steady state equations are solved simultaneously and terms are collected, a simple quadratic must be solved. From the general mass balance equation:

rate of change = input - output + reaction

the cell concentration in the ith unit is:

$$\frac{dX_i}{dt} = \frac{F_i Xo_i}{V_i} + \frac{\omega F_i C_i X_i}{V_i} - \frac{(1+\omega) F_i X_i}{V_i} + \mu X_i \quad (4.17)$$

Let $D = \dfrac{F_i}{V_i}$

$$D\,Xo + \omega D C X - (1+\omega) D X + \mu X = 0 \quad (4.18)$$

where Xo = concentration of cells in feed
 ω = recycle ratio
 C = concentration of cells in recycle stream

An analogous derivation for S gives:

$$D\,So + \omega D S - (1+\omega) D S - \mu X/Y - m X = 0 \quad (4.19)$$

or $\quad D\,So - D S - \mu X/Y - m X = 0 \quad (4.20)$

where m = the maintenance energy
 Y = the yield constant

The notation (i) has been omitted for brevity. Using the Monod equation to relate μ and S, and solving Equation 4.18 for X:

$$X = \frac{-D\,Xo}{\omega D C - (1+\omega) D + \hat{\mu}} \quad (4.21)$$

Substituting into Equation 4.20:

$$[\,D(1+\omega)Y - D\omega YC - \hat{\mu}Y\,]\,S^2 +$$
$$[\,D\omega YCSo - D(1+\omega)YSo + \hat{\mu}YSo - D\omega YCK_s +$$
$$D(1+\omega)YK_s + mYXo + \hat{\mu}Xo\,]\,S +$$
$$D\omega YCK_s So - D(1+\omega)YK_s So + mYK_s Xo = 0 \quad (4.22)$$

Now, let

$A = D(1+\omega)Y - D\omega YC - \hat{\mu}Y$

$B = D\omega YCSo - D(1+\omega)YSo + \hat{\mu}YSo - D\omega YCK_s +$
$\qquad D(1+\omega)YK_s + mYXo + \hat{\mu}Xo$

$$C = D\omega Y C K_s S_o - D(1 + \omega) Y K_s S_o + m Y K_s X_o$$

Equation 4.21 can be solved by the quadratic equation as:

$$S = \frac{-B \pm \sqrt{B^2 - 4AC}}{2A}$$

The impossible solution is discarded. Knowing S, it is simple to solve for X:

$$X = \frac{D(S_o - S)}{\mu/Y + m} \tag{4.23}$$

Cell mass balance around the separation step gives:

$$(1 + \omega)FX = FXE + \omega FCX \tag{4.24}$$

$$XE = \frac{(1 + \omega)FX - \omega FCX}{F} \tag{4.25}$$

$$= (1 + \omega - \omega C) X \tag{4.26}$$

and substrate balance means that:

$$SE = S \tag{4.27}$$

The values of XE and SE are the S and X terms for the subsequent stage. The productivity in any stage is merely the product of the dilution rate and the cell concentration in the exit stream.

With the overall dilution rate based on the total volume of all the stages (denoted D_o):

$$D_o = F_o (1 + R_1 + \ldots + R_n) \tag{4.28}$$

Where Fo is the rate of the initial feed stream and the R's are ratios of additional feeds to Fo.

The solution that is not physically impossible is exact for the specified system, but some of the assumptions may not be rigorous. For example, a constant yield coefficient is assumed although it it known that starving cells at very low dilution rate differ from rapidly dividing cells. Further research with continuous cultivation of microorganisms is needed to provide relationships on which to base improved computer analysis.

EXECUTION OF THE PROGRAM

The program listing is available from its authors on request. There is no BASIC version, but two preliminary BASIC programs used for recyle and two-stage continuous fermentation are in Appendix 4. The program "IGPMSC" (Interactive Graphics Program for Multistage Continuous Culture) has prompts so that the instruction manual is consulted very infrequently. Up to four stages of continuous culture can be requested, and this is an arbitrary restraint so that the display fits the screen of the terminal nicely. The manual is available in a file named "INFO".

The layout of the system is drawn and the specifications are echoed as shown in Fig. 4-4. Feed streams and recycle loops are drawn only if specified.

Almost all of the interaction with the computer is performed with the light pen, and typing is only needed for numbers. The menu shown at the side in Fig. 4-4 interacts with a light pen for easy selection of options. These are:

QUIT -- causes immediate termination.

HDCOPY -- this will generate a paper copy of the current screen contents; when this command is selected the message 'HARDCOPY ACTIVE' will appear on the upper right corner of the screen, on successful copy the message will disappear and the program continues.

RECALL -- this will cause the program to recall the original graphs of results.

These options are common to both parts of the program. (Fig.4-4 and Fig.4-5)

RESET -- reinitializes the program for a fresh start.

CHANGE -- this allows user to change parameters in individual stages. After all the changes have been made, the program will draw a new block diagram.

RUN -- this will cause the program to proceed with the analysis and produce output plots.

When specifications are altered, the screen is cleared and prompts appear for the needed information. When the analysis option is selected, the program calculates the results extremely rapidly but there is a brief wait because the graphical display routines are rather slow. A typical result is shown in Fig. 4-5.

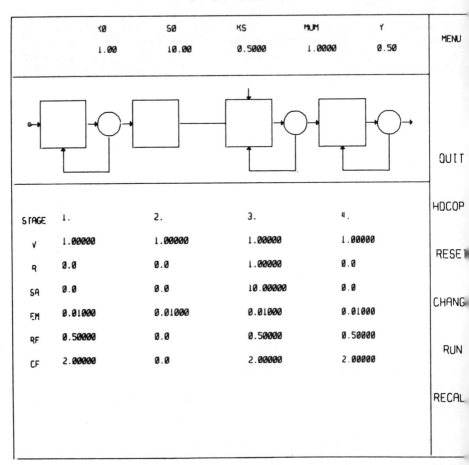

Figure 4-4

LAYOUT FOR MULTI-STAGE GAME

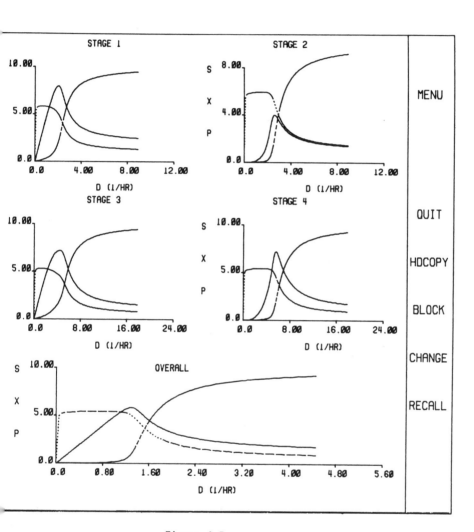

Figure 4-5

RESULTS FOR MULTI-STAGE GAME

Again, the menu in Fig. 4-5 with light pen gives easy selection of options. These are:

BLOCK -- the program returns to the block diagram as shown in Fig. 4-4.

CHANGE -- this will allow the user to change parameters temporarily, upon selection of BLOCK the program will return to the original block diagram without carrying any changes made during the CHANGE option.

The results are graphed with organism concentration, substrate concentration, and productivity (organism concentration times dilution rate) versus dilution rate. In addition to graphs for each stage, there is a graph of overall performance determined from outputs from the final stage and overall volumes and feeds. This overall graph facilitates comparisons between various configurations, but the individual stages are important in explaining the results and devising improved configurations.

As is well known, recycle of cells permits the dilution rate to exceed the maximum specific growth rate without washout. In a cascade of continuous fermenters, it may be desirable to make use of recycle in some stages while employing other stages for exhaustion of substrate. In stages where cells are starving, the biological coefficients may differ from those in other stages. The present algorithm does not adjust parameters automatically, and the user must decide on appropriate constants. Because substrate level changes with dilution rate and cells may encounter starvation, it would be of benefit to have biological constants be functions of other variables. Unfortunately, understanding of metabolism is not yet sufficiently complete to develop solid relationships for this program.

A detailed discussion of the many configurations that have been tested would far exceed the space allocated to this chapter. However, these experiments at the computer terminal are interesting and permit testing of original concepts. Additional substrate feed in later stages does not increase the overall maximum dilution rate in a single-culture, single-substrate system; but it has promise in a multi-stage multi-substrate system.

REFERENCES

H.R. Bungay, L.S. Clesceri, and N.A. Andrianas, (1981), "Autoselection of Very Rapidly Growing Microorganisms", Advances in Biotechnology, Proceedings of International Fermentation Symposium, London, Ont., ed. M. Moo-Young

Malek, I., and Z. Fencl, (1966), "Theoretical and Methodological Basis of Continuous Culture of Microorganisms", Academic Press

Bailey, J.E., and D.F.Ollis, (1977), "Biochemical Engineering Fundamentals", McGraw-Hill

Lam, H.L.Y., and H.R. Bungay, (1983), "Interactive Graphics Simulation of Multistage Continuous Culture", in Engineering Images for the Future", ed. L.P. Grayson and J.M. Biedenbach, Proc. 1983 Ann. Conf. Am. Soc. Engr. Educ., 2: 474-494

Chapter 5

MIXED CULTURES

Of the numerous situations where mixed cultures are important, microbiological treatment of wastes is particularly interesting because very complicated mixed cultures interact in many ways. Such processes have elective cultures in which the proportions of different species can shift dramatically in response to changing nutrition or physiological conditions. There are a some industrial or food fermentations that use mixed cultures. A second organism sometimes is inoculated into food and beverage fermentations to affect taste or texture. A rare case of a synergistic industrial fermentation is the production of beta-carotene with plus and minus strains of Blakelea trispora. This could be common in the future when microbial metabolism is better understood and situations are devised in which mixed cultures can help each other. Mixtures of microorganisms are employed in the processes shown in Table 5.1.

There is an interesting area of research on defined mixtures of microorganisms, and they behave much differently in combination than alone. Definitions of various interactions in Table 5.2 are difficult to apply to real systems where there is highly complicated interplay among organisms having various roles with respect to each other.

Mathematical models can be derived for any mixed culture system, but usually little is known about how to formulate the interaction terms. This highlights a gap in knowledge where research can be justified easily.

COMPETITION

Assuming that specific growth rate coefficients are different, no steady state can be reached in a well-mixed continuous culture with competing types present because if one were at steady state with its specific growth rate coefficient equal to D, another faster or slower organism

TABLE 5.1
MIXED CULTURE PROCESSES

PROCESS	TYPES OF ORGANISMS
Commercial	
Alcoholic beverages	Various yeasts, molds, bacteria.
Sauerkraut	L. plantarum, other bacteria.
Pickles	L. plantarum, other bacteria.
Cheeses	Propionibacteria, molds; possibly many other microorganisms.
Lactic acid	Two lactobacillus species.
Beta-carotene	Two sexes of Blakelea trispora.
Waste treatment	
Trickling filters	Zoogloea, protozoa, algae, fungi.
Activated sludge	Zoogloea, Sphaerotilus, yeasts, molds, protozoa
Sludge digestion	Cellulolytic, acid-forming and methanogenic bacteria,
Sewage lagoons	Many types; many families.

TABLE 5.2
SOME DEFINITIONS OF MICROBIAL INTERACTIONS

Competition	A race for nutrients and space
Predation	One feeds on another
Commensalism	One lives off another with negligible help or harm
Mutualism	Each benefits the other
Synergism	Cooperative metabolism
Antibiosis	One excretes factor harmful to other

could not also have its specific growth rate coefficient equal to D. With a derivative unequal to zero, there is no steady state. In other words, when the fast organisms reach steady state, the slower organisms have a negative derivative and must continually decrease. The net effect is that the faster-growing type takes over while the other declines to zero. Theory does not seem to hold for natural ecosystems. In real systems, even those that approximate well-mixed continuous cultures, there may be profound changes in relative numbers of the various organisms present, but complete take over by one type is extremely uncommon. It is likely that some species persist by attaching to surfaces to resist washout, and others may be coupled nutritionally so that their contributions are needed. For example, when the main carbohydrate is cellulose, slow-growing producers of cellulase enzymes may be essential for releasing sugar for other organisms that may grow rapidly. In fact, an uncommon substrate should insure the persistence of the one type of organism able to metabolize it. Survival of a broad range of species is highly advantageous in natural systems because a needed type will be present when unusual conditions or nutrients are encountered.

Competition in a chemostat is modeled with the equations:

$$dA/dt = \mu_a A - D A \tag{5.1}$$

$$dB/dt = \mu_b B - D B \tag{5.2}$$

$$dS/dt = D So - D S - \mu_a A/Y_a - \mu_b B/Y_b \tag{5.3}$$

where
A = concentration of one type of organisms
B = concentration of another type
S = concentration of limiting nutrient for each
So = substrate concentration in feed
μ_a = specific growth rate coefficient for A
μ_b = specific growth rate coefficient for B
Y_a = yield coefficient for A
Y_b = yield coefficient for B

and there are separate Monod equations for A and B.

A simple case of a culture of a slower organism at steady state suddenly contaminated by a faster growing organism is shown in Fig. 5-1. The simulation was with SIM4, and the relevant sections of a SIMBAS program for this simulation are in Fig. 5-2. Note that the initial conditions specified for this system did not respond to steady state, and the substrate and cell concentrations were still drifting when the fast-growing cells were introduced. This was not corrected because it demonstrates a type of error. One way of establishing steady state would have been to wait longer before inoculating the faster cells.

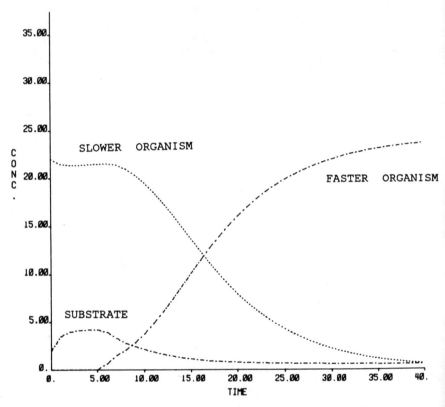

Figure 5-1
CONTAMINATION BY FASTER-GROWING ORGANISMS
(System not Initially at Steady State)

```
1 REM CHEMOSTAT CONTAMINATED AT 5 HR.
20 DIM I(20),O(20),TO(20),T6(20)
30 D=.2
32 F=25
40 REM D=DILUTION RATE, M=SLOWER GROWTH COEFF.
42 REM G=FASTER GROWTH COEFF., 1 IS SLOW ORGANISM
44 REM 2 IS FAST ORGANISM, 3 IS SUGAR
46 PRINT "TIME        SLOW ORG.       FAST ORG.           SUGAR"
51 PRINT
56 T1=.2
57 T2=50
58 T3=1
80 N=3
82 O(1)=10
84 O(2)=0
86 O(3)=3
1000 I(1)=O(1)*(M-D)
1010 I(2)=O(2)*(G-D)
1020 I(3)=D*(F-O(3))-2.1*M*O(1)-2.05*G*O(2)
1030 M=.3*O(3)/(2+O(3))
1040 G=.8*O(3)/(1.5+O(3))
1999 GOTO 100
2000 PRINT T,O(1),O(2),O(3)
2020 IF T=5 THEN T6(2)=1
2045 RETURN
```

Figure 5-2

RELEVANT PORTIONS OF SIMBAS PROGRAM FOR COMPETITION

CONCENTRATION-COUPLED CONTINUOUS CULTURE

A continuous culture with feed back control of nutrient concentration can select rapidly dividing cultures. As the cells call for increased pumping of nutrients to hold the concentration constant, there is an accompanying decrease in residence time because the culture volume is fixed. Various cells are competing, and those that do not double during their residence time are washed out with no progeny to contribute to the mixed population. It is a powerful advantage for cells that divide more rapidly to persist in the reactor as slower cells tend to be lost. After several days of competition, very rapidly growing cultures are selected (Bungay, et al, 1981).

Mass balance equations for the case with the substrate concentration fixed are the same as derived for the mathematics of competition. Now the only derivative that can be set to zero is dS/dt, and Equation 5.3 can be solved for D to yield:

$$D = \frac{\mu_a A + \mu_b B}{Y(So - S)} \qquad (5.4)$$

Instead of letting A and B stand for two different cultures, B could be a faster-growing mutant of A. It is quite easy to handle these equations with SIMBAS, but the rational for mutation is not easily stated mathematically. Mutation is a random event, so we might employ a random number generator to change the values for the specific growth rate coefficients. When to change them and how much the changes should be are mostly guesswork at our present state of knowledge.

INOCULATION OF A SLOWER CULTURE TO A CHEMOSTAT

There are several exceptions to take over by the faster organism for competition in continuous culture. Nutritional interdependencies, wall attachment, or some other special circumstances can allow the slow organisms to persist. When one culture attaches to surfaces to resist washout, there can be steady states with a faster, freely suspended organism, or substrate level can determine take over (Baltzis and Fredrickson, 1983). Another, closely related system that we have been exploring is continuous inoculation of the slower organism so that its washout is impossible (Bungay, 1984). Computer simulation of a chemostat of rapidly growing organisms shows that slower growing organisms can dominate if inoculated continuously. The fraction of slower organisms to be fed to the chemostat for dominance depends on the Monod coefficients. Simulation shows that this technique works well except when the desired culture grows at less than half the rate of the organisms that are to be replaced.

Practical reasons for controlling domination in a chemostat are :
1. minimizing the hazard of contamination.
2. controlling proportions in a synergistic fermentation.
3. creating a possibility of non-aseptic operation.

Although continuous fermentation is the norm for biological treatment of wastes, it is not very common for industrial fermentations. If means of reducing contamination and lowering costs by using simpler equipment could be found, there might be a more rapid switch to continuous processes. If a small continuous fermenter were used to grow these very rapidly dividing organisms and used to inoculate a production-size continuous fermenter, it is very unlikely that contamination would occur. Considering that a slow culture can dominate if inoculated continuously, a logical advance is to inoculate a very fast culture that would displace potential contaminants. While the inoculum vessel for vigorous growth of the desired culture may be operated aseptically, the main vessel should perform well with few precautions. Such a system would represent a radical departure from current fermentation practice.

TWO-CULTURE CHEMOSTAT WITH INOCULATION

The statement of the problem is:
a fast growing organism, A, is present initially,
a slower-growing organism, B, is inoculated continuously,
find the concentration of B when A is present,
determine the range of dilution rates where A washes out,
and find the concentration of B when A is not present.

As developed earlier, if steady state is reached with A present, μ_a = D, and the equations can be solved for A and S at any specified D. Note that the A equation will not have any additional terms when a second organism, B, is also in the continuous culture, thus μ_a will still equal D whenever steady state is reached with A present. This is a simple but profound concept that make analysis of the mixed culture quite easy. When B is added continuously, the B equation becomes:

$$dB/dt = D\, I_b - D\, B + \mu_b\, B \qquad (5.5)$$

and a Monod equation for B is:

$$\mu_b = \hat{\mu}_b\, S / (K_b + S) \qquad (5.6)$$

where B = concentration of second organism
μ_b = specific growth rate coefficient of #2
$\hat{\mu}_b$ = maximum coefficient for #2
K_b = Monod coefficient for #2
I_b = concentration of B in feed stream

When D is specified, S can be calculated from the Monod equation for organism A in the form :

$$S = D\, K / (\hat{\mu}_a - D) \qquad (5.7)$$

Because the known value of S defines the rate coefficient in Monod equation #2, the dB/dt equation can be solved knowing I_b to give :

$$B = D\, I_b / (D - \mu_b) \qquad (5.8)$$

With S and B known, A can be calculated using a rearranged form of the S equation :

$$A = Y\, So - D\, S\, Y - Y\, \mu_b\, B / (D\, Y_b) \qquad (5.9)$$

In other words, μ_a and S are fixed by the property that steady state is reached with A present, and B and A are calculated from additional mass balance equations.

A simple computer algorithm was devised and programmed for the RPI Center for Interactive Computer Graphics. The program is in Appendix 4. The dilution rate is incremented in steps from zero to slightly more than the maximum specific growth rate coefficient for the faster organism. A typical simulation is shown in Figure 5-3. where A is assumed to grow faster than B. Feeding B causes A to wash out before D equals the maximum growth rate coefficient for A. When there is a small difference in specific growth rates, only a slight amount of B in the feed stream will cause A to wash out at moderate dilution rates. On the other hand, when A grows much faster than B, B can predominate only if fed at relatively high concentrations.

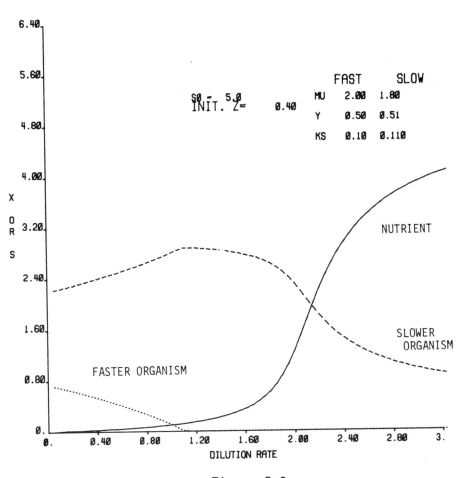

Figure 5-3
CHEMOSTAT WITH INOCULATION OF SLOWER ORGANISM

Another condition that is possible at steady state is that A = 0. With no A present, it is possible to calculate B, S, and μ_b from three simultaneous equation - the B equation, the S equation, and the second Monod equation. This leads to the algebra shown below which simplifies to a quadratic equation in S where only one solution has physical meaning. Of course, when S is known, B is easily determined. Endogenous metabolism was not considered, but the graphs of B and A versus D would be affected only at low dilution rates when reasonable values of maintenance energy coefficients are tested.

Let us consider the conditions at which A cannot barely survive. Equation 5.8 set to zero and with Y factored out gives:

$$S_o - DS - \mu_b B / (D Y_b) = 0 \tag{5.10}$$

or multiplying through by D Yb:

$$S_o D Y_b - D^2 S Y_b - \mu_b B = 0 \tag{5.11}$$

Substituting Equation 5.8 for B and Equation 5.6 for μ_b, simplifying, and collecting terms, there is an equation of the form:

$$A S^2 + B S + C = 0$$

where $A = D^2 Y_b \hat{\mu}_b - D^2 Y_b$

$B = S_o D Y_b - S_o Y_b \hat{\mu}_b - D^2 Y_b K_b - D I_b$

$C = S_o D Y_b K_b$

$$S = \frac{-B \pm \sqrt{B^2 - 4AC}}{2A}$$

A program for this calculation is left as an exercise for the reader. This simulation is useful for testing permutations of coefficients to find the minimum dilution rates that cause washout of the faster organisms. The results help in planning experiments to verify the theory. However, the analysis is also valuable in educating students, and the simulation is used as an assignment in our course in Advanced Biochemical Engineering.

PREY-PREDATOR SYSTEMS

The classic Lotka-Volterra equations for predation are based on the concept of encounters between prey and predators. As in mass action for chemical reactions, the rate of biological events is taken as proportional to the product of the concentrations of the participating species. These events are encounters that let predators kill their prey, and the predators use this feeding to multiply. Equations based on simple mass balances for continuous culture are:

$$dV/dt = \mu_V V - D V - K1 V P \qquad (5.12)$$

$$dP/dt = K2 V P - D P \qquad (5.13)$$

where
- V = concentration of victims, (hosts, prey)
- P = concentration of predators
- μ_V = specific growth rate coefficient for V
- D = dilution rate
- $K1$ = killing coefficient
- $K2$ = efficiency of killings in producing new predators

The effectiveness of collisions in destroying victims determines K1, and the growth of predators also relates to collisions through K2. Some simulations of this system are shown in Fig. 5-4, and the relevant SIMBAS sections are in Fig. 5-5. Real systems do have this type of oscillating behavior, but frequencies and amplitudes are erratic.

Ratnam, Pavlou, and Fredrickson (1982) showed that wall attachment attenuates the extremes of population during oscillations as compared to systems where wall attachment was minimized by coating the vessel walls with a silicone material.

IMPROVED MODELS FOR PREDATION

Several authors have pointed out limitations in the Lotka-Volterra equations. A typical reference is Bungay and Paynter (1972). The principal problem is the growth rate of predators because the equations indicate no limit. That is to say, if the concentrations of victims and predators are both large, the rate of predator increase could be impossibly great. Of course, all living things have their maximum growth rate, and most predatory species do not grow at very high rates. By analogy to other biological growth rate equations, the growth rate of predators should be a function of the concentration of their food, the victim concentration. The predator mass balance equation would be:

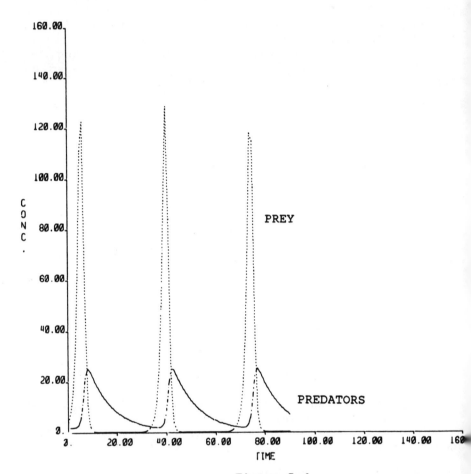

Figure 5-4
PREDATION SIMULATION WITH LOTKA-VOLTERRA MODEL

```
20 REM LOTKA-VOLTERRA MODEL
30 D=.25 :REM DILUTION RATE
32 MU=1.05 :REM CONSTANT GROWTH RATE FOR PREY
51 PRINT
52 PRINT"TIME    A    B              PLOT OF B"
54 K1=.3 :REM KILLING EFFICIENCY
55 K2=.03 :REM PREDATOR GROWTH EFFICIENCY
56 LET T1=.2
57 T2=100
58 LET T3=1!
80 LET N=2
86 O(1)=50 :REM PREY
88 O(2)=5 :REM PREDATORS
1000 I(1)=MU*O(1)-D*O(1)-K1*O(1)*O(2)
1010 I(2)=K2*O(1)*O(2)-D*O(2)
1999 GOTO 100
2000 LET B=25+O(2)*3
2010 PRINT T;O(1);O(2);TAB(B);"*"
2045 RETURN
```

Figure 5-5

PORTIONS OF SIMBAS FOR LOTKA-VOLTERRA PREDATION

$$dP/dt = \mu_p P - D P \qquad (5.14)$$

where μ_p = the specific growth rate coefficient for P, and

$$\mu_p = \hat{\mu}_p V / (K_v + V) \qquad (5.15)$$

The μ_p equation is assumed to have a Monod form, and K_v is some constant.

Another questionable assumption is that the victim growth rate is constant. In microbial systems, the growth rate is a function of limiting substrate concentration. The mass balance for victims should be:

$$dV/dt = \mu_v V - D V - K1 V P \qquad (5.16)$$

and there must be a mass balance equation for substrate:

$$dS/dt = D So - D S - \mu_v V / Y \qquad (5.17)$$

where the S equation mimics Equation 5.3.

Note that K2 times V in Equation 5.13 is replaced by μ_p. In other words, the Lotka-Volterra derivation is equivalent to a linear relationship of μ_p to V while the Monod relationship is linear at low concentrations of V and approaches a maximum at high concentrations of V.

Simulation of a prey-predator system with only collision theory for the predator growth rate but with a Monod equation for the growth rate of prey resulted in Fig. 5-6. It took considerable tinkering with the coefficients to get oscillating populations, and the figure suggests that this oscillation would dampen out at extended times. The original simulation with Lotka-Volterra assumptions had a rapid increase in prey concentration to help oscillations. With the prey dependent on a limiting nutrient, their concentration is restricted so that large numbers do not unbalance the system.

Some simulations that incorporate both a substrate limitation for V and a victim limitation for P are shown in Fig. 5-7. Relevant sections of a SIMBAS program are in Fig. 5-8. The reader certainly should be able to develop these simulations unaided, but some tinkering is required to find coefficients that produce reasonable answers.

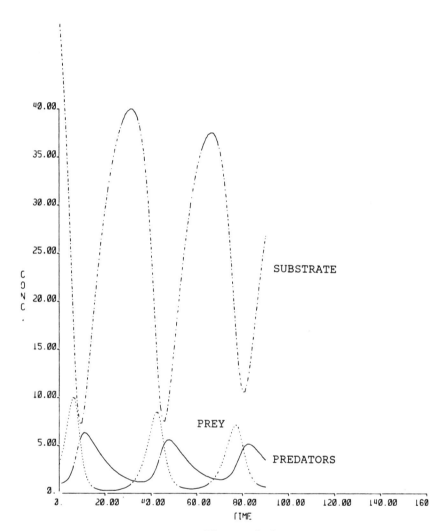

Figure 5-6
PREDATION WITH PREY HAVING NUTRIENT LIMITATION

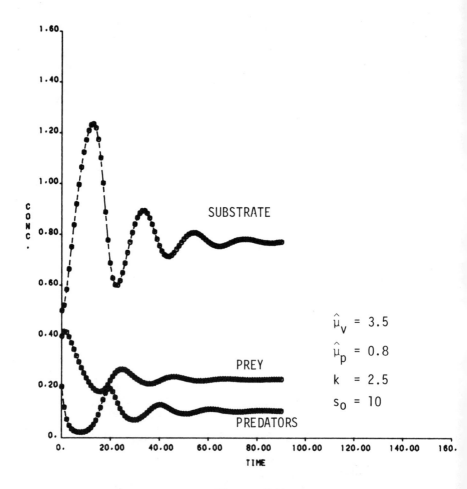

Figure 5-7

SIMULATION OF PREDATION WITH MONOD FUNCTIONS

```
10 REM PREY-PREDATOR WITH MONOD RELATIONSHIPS
20 DIM I(20),O(20),T0(20),T6(20)
30 D=.25
32 MX=1.2
34 KS=4
36 PX=.5
38 KP=2
40 Y=.55
42 F=100
50 PRINT"TIME    PREY    PREDATORS    MUV        MUP
51 PRINT
54 K1=.3
55 K2=.03
56 T1=.2
57 T2=200
58 T3=1
80 N=3
86 O(1)=5
87 O(3)=5
88 O(2)=1
1000 I(1)=MU*O(1)-D*O(1)-MP*O(2)
1005 IF O(1)<0 THEN O(1)=.01
1010 I(2)=MP*O(2)-D*O(2)
1020 I(3)=D*(F-O(3))-MU*O(1)/Y
1025 IF O(3)<0 THEN O(3)=.1
1030 MU=MX*O(3)/(KS+O(3))
1040 MP=PX*O(1)/(KP+O(1))
1999 GOTO 100
2000 LET B=25+O(2)/3
2010 PRINT T;O(1);O(2);O(3);MU;MP
2045 RETURN
```

Figure 5-8

PORTION OF SIMBAS FOR PREDATION, TWO MONOD FUNCTIONS

Problems in modeling protozoal systems in continuous culture are covered nicely by Pavlou and Fredrickson (1983). They note that prey-predator systems with protozoa feeding on bacteria may oscillate at the start of an experiment, but the oscillations tend to dampen out. In addition to such effects as wall attachment and non-homogeneous distribution, they postulated another complication - inability of the protozoa to collect bacteria over the whole range of possible sizes. When this was incorporated into a mathematical model, there was good agreement with actual data on the time course of populations of microorganisms. When the bacteria are growing, there is a tendency to oscillations. The populations tend to reach a steady state when the protozoa are fed non-growing bacteria.

The improved equations also have some faults. The most serious may be the instantaneous adjustment of growth rates. Mateles, Ryu, and Yasuda (1965) were the first to show that there can be some rapid adjustment of growth rate to a step up in substrate concentration, but time is required for a full adjustment. By analogy, if cellular biochemistry is viewed as an assembly line, there may be some excess capacity available for a sudden increase in input units to the line, but more assembly lines must be built to handle a large increase in input. Biochemical assembly lines involve nucleic acids, genetic controls, protein synthesis, and the like which take time to build. There can be a considerable wait before full adjustment to a step up in substrate concentration is over. A step down underutilizes the assembly line but necessitates no delay.

There are many ways to incorporate delay into simulation of differential equations. One simple method is to set up an array of values and to step through the array adding the new value to the head of the list while removing the value that has been stored for the desired time from the bottom of the list. A logic diagram for such a delay is shown in Fig. 5-9. It would not be difficult to devise a program to delay part of a step and to have immediate response to the rest.

A simple BASIC program for delay is shown in Fig. 5-10. However, this is a clumsy method because the array must be reshuffled each time a number is placed at the top, numbers are moved, and a number is taken away. A much more efficient method is to move the subscript of the array in a circular fashion. When the subscript reaches the end, it goes back to the beginning. The number to be delayed is added at the value of the subscript, and the old, delayed value is extracted at the subscript plus one. A typical BASIC program for this method is given in Fig. 5-11.

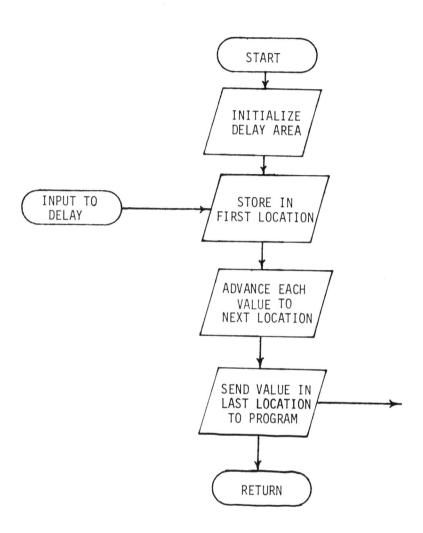

Figure 5-9

LOGIC FOR INFERIOR SUBROUTINE FOR DELAY

```
10 REM DELAY FOR 25 STEPS
20 DIM D(30)
30 PRINT
40 PRINT "TIME VALUE IN VALUE OUT"
50 T=0
60 T=T+1
70 D(27)=0
80 IF T>5 THEN D(27)=T-5
90 FOR I=1 TO 26
100 D(I)=D(I+1)
110 NEXT I
120 PRINT T, D(27), D(1)
130 IF T<35 THEN 60
```

Figure 5-10

DELAY WITH INFERIOR ALGORITHM

```
10 REM IMPROVED DELAY EXAMPLE
20 REM INDEX MOVES, NOT NUMBERS
30 REM I IS INPUT VALUE, D IS DELAY SIZE
40 DIM W(50)
50 INPUT "DELAY SIZE = ";D
60 FOR J=1 TO 50
70 W(J)=0
80 NEXT J
90 T=0
100 PRINT
110 PRINT "TIME VALUE IN VALUE OUT"
120 T=T+1
130 I=0
140 IF T>5 THEN I=T-5
150 J=J+1
160 IF J>D THEN J=1
170 N=J+1
180 W(J)=I
190 IF J=D THEN N=1
200 PRINT T,I,W(N)
210 GOTO 120
```

Figure 5-11

DELAY WITH IMPROVED ALGORITHM

The delay of a value for a period of time is called distance-velocity lag. Many flow processes exhibit this lag when there is plug flow. For example, passage of a solution through a pipe may have negligible mixing or dispersion, thus an element of fluid may be assumed to traverse the pipe unchanged. A graph of the inlet concentration will be exactly the same as a graph of the outlet concentration except for a displacement in time.

Care must be exercised in incorporating time delay into a SIMBAS or SIM4 simulation because the integration step size is not the same as the number of calculations. These programs employ a fourth-order Runge-Kutta integration scheme, so four calculations take place during each time interval. The delay routine should be with the section for differential equations and should account for the step size times four. For example, if the desired delay is 5 time units and the integration step is 0.1, the array size for delay is 5 x 10 x 4 = 200.

FIRST-ORDER DELAY

A distinctly different type of delay or lag results from the response time of process elements. This will be explained in more detail in Chapter 8 as part of dynamic analysis. A microbial growth rate might start adjustment to a step increase in substrate concentration immediately and approach its final value asmyptotically. This type of response is shown in Fig. 5-12. Not enough is known about the true control system for microbial growth rate to decide whether distance-velocity lag or first-order delay is more appropriate for modeling. In the author's experience, either type of delay seems to improve a simulation markedly with respect to matching real data, but neither is significantly better than the other.

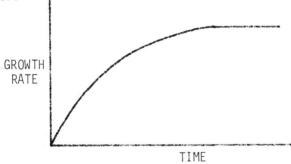

Figure 5-12
RESPONSE WITH FIRST-ORDER DELAY

First-order delay is modeled with one additional differential equation. The rate of response is proportional to the driving force, and this equals the difference between the present value and the final value. The equation is:

$$dR/dt = K (Rf - R) \qquad (5.18)$$

where R = the value of the variable
Rf = the value being approached
K = a constant

Higher order processes are common in many systems or processes, and there have been some simulation studies on microbial systems with consideration of higher order effects, but the assumptions are too speculative to justify coverage in this book.

ECOSYSTEM THEORY

An entire ecosystem is much more complicated than the simple microbial interactions that have been described here. A large group of modelers with a highly theoretical perspective is emphasizing spatial order, probability, theory of fitness of mutational changes to the restraints of the environment, events at the molecular level, amino acid incorporations into proteins, and the like. While this work is important and should not be dismissed casually, it is too far afield from the elementary, more applied material in this book. Some typical references are:

M. Conrad, (1982) "Natural Selection and the Evolution of Neutralism", BioSystems, 15: 83-85

M. Conrad, (1983) "Adaptability", Plenum

L.L. Gatlin, (1972) "Information Theory and Living Systems", Columbia Univ. Press

R.C. Lewontin, (1974) "The Genetic Basis of Evolutionary Change", Columbia Univ. Press

M. Volkenstein, (1979) "Mutations and the Value of Information", J. Theor. Biol. 80: 155-169

Wimpenny, J.W.T., (1981) "Spatial Order in Microbial Systems - Review" Biol. Rev. 56: 295-342

REFERENCES

Baltzis, B.C., and A.G. Fredrickson, (1983) "Competition of Two Microbial Populations for a Single Resource in a Chemostat when One of Them Exhibits Wall Attachment" Biotechnol. Bioengr. 25: 2419-2439

Bungay, H.R., and M.J.B. Paynter, (1972) "Models for Multi-component Microbial Ecosystems", in Water-1971, A.I.Ch.E. Symp. Ser. 124 68: 20

Bungay, H.R., (1976) "Formulating Predator Growth Rate Terms for Population Models", in Modeling Biochemical Process in Aquatic Ecosystems, ed. R.P. Canale, Ann Arbor Sci. pp. 377-382

Bungay, H.R., (1984) "Population Control in Continuous Fermentation" in Frontiers in Chemical Reaction Engineering, Vol. 1, ed L.K. Doriaswamy and R.A. Mashelkar, Wiley International

Bungay, H.R., L.S. Clesceri, and N.A. Andrianas, (1981) "Auto-selection of Very Rapidly Growing Microorganisms", in Advances in Biotechnology, ed. Moo-Young, C. Vezina, and C.R. Robinson, Vol. 1: 235-241

Mateles, R.I., D.Y. Ryu, and T. Yasuda, (1965) "Measurement of Unsteady State Growth Rates of Microorganisms", Nature 208: 263-265

Pavlou, S. and A.G. Fredrickson, (1983) "Effects of the Inability of Suspension-Feeding Protozoa to Collect all Cell Sizes of a Bacterial Population", Biotech. Bioengr. 25: 1747-1772

Ratnam, D.A., S. Pavlou, and A.G. Fredrickson, (1982) Biotechnol. Bioengr. 24: 2675

ADDITIONAL READING

A.T. Ball and J.H. Slater, ed. "Microbial Interactions and Communities", Academic Press (1982)

Chapter 6

ENVIRONMENTAL SYSTEMS

Lakes, rivers, impoundments, and other water systems are of interest to biochemical engineering as sources for process waters and cooling water and as sinks for plant effluents. Although the environmental engineers take prime responsibility for water treatment and waste treatment, it is worthwhile to present some simulations directed toward water and wastewater.

THE DISSOLVED OXYGEN SAG CURVE

Dissolved oxygen is the key parameter in most rivers because the biological community is seriously damaged by inadequate oxygen. Fish suffocate and float on the surface to indicate that the stream is sick. The fish species that survive at lower D.O. are carp and the like instead of desirable game fish such as trout. If much of the stream becomes anaerobic, the water may take on a foul appearance and may emit unpleasant odors. A traditional index of pollution is B.O.D., biochemical oxygen demand. This expresses organic material as the equivalent of oxygen consumed through its metabolism by microorganisms. Although there is a trend to specific measurements of individual pollutants, B.O.D. is still the most used term for organic pollutants.

The Streeter-Phelps equations are the classic model for decay of pollution and consumption of oxygen in a stream:

$$dL/dt = - K1\ L \qquad (6.1)$$

$$dD/dt = K1\ L - K2\ D \qquad (6.2)$$

where L   = the concentration of pollutants (B.O.D.)
       D   = oxygen deficit (concentration of oxygen if saturated with air minus the actual concentration.)
       K1   = coefficient for assimilation of pollution,
       K2   = reaeration coefficient
       t   = time, but distance can be substituted because distance equals flow times time.

Equation 6.1 implies that the disappearance of pollution is first-order. In earlier chapters, the concentration of limiting nutrient had roughly an S-shaped plot versus time. Much lower initial concentrations of nutrients are usually encountered in rivers and streams, and the biological community may not be growing rapidly as in fermentation systems. In fact, organic materials may be following various time behaviors of their concentrations. A gross measure of all metabolizable ingredients as B.O.D. often fits a first-order model. In other words, the net effect for many different compounds may appear to have decreasing slope so that the logarithm gives a linear plot. This is logical because easily degraded materials give a steep initial slope while recalcitrant materials give a very low rate.

Reaeration of the stream is assumed to be directly proportional to the driving force which is the displacement from the air-saturation value. Thus Equation 6.2 is based on stoichiometric consumption of oxygen for the metabolism of L with reaeration at a rate proportional to the deficit. The equations have exactly the same form as those in the SIMBAS example given in Chapter 2. However, it is customary to plot oxygen concentration instead of deficit. A typical plot using the demonstration example in SIMBAS but subtracting D from the saturation value of oxygen is shown in Fig. 6-1.

In a real stream, the coefficients may change because of rapids, widening of the river, deep sections, or the like. Using IF statements to alter coefficients gave the graph in Fig. 6-2. Similar logic could account for adding slugs of pollution along the stream, but SIMBAS and SIM4 have a complication. Because of the fourth-order Runge-Kutta integration routine, intermediate values of variables are stored in the computer. A new specification of the output of an integration causes it to be averaged with the stored value, but the new pollution should be additive not averaged. The desired change can be implemented by changing I(j) where j stands for the equation of interest,

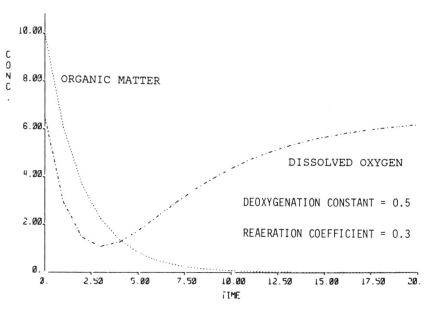

Figure 6-1
DISSOLVED OXYGEN SAG CURVE

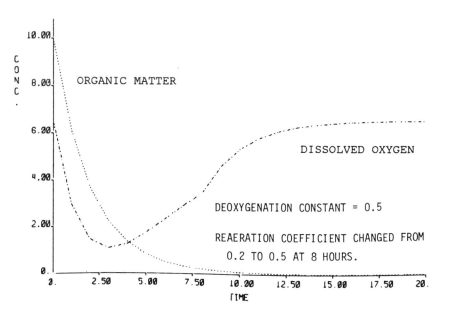

Figure 6-2
OXYGEN SAG CURVE WITH CHANGE IN AERATION AT 8 HOURS

and also equating T6(j) which is the stored value. This was also explained in Chapter 3 for the simulation of back seeding.

Although there are simple analytical solutions of the Streeter-Phelps equations, computer simulation is easy and produces graphs as well as numbers. Repeating the calculations over and over to develop a series of points is essentially the same as using a simulation package but it less straight forward. More elaborate and sophisticated equations for modeling streams usually must be handled by numerical techniques with the computer because there is no direct solution.

STREAM POPULATION DYNAMICS GAME

Biological indicators of stream pollution tend to be more valid than chemical analysis because chemicals can be added intermittently while organisms reflect cumulative effects. If sampling for chemical assay misses the times when chemicals were flushed to the stream, a false impression of the amount of pollution is given. In other words, analysis of a grab sample gives a snapshot of stream chemistry with little indication of what may have happened earlier. The biological community has been exposed over a period of time and reflects any insults. Enumerating organisms of several species can be highly informative for characterizing damage from pollution, but only if a trained investigator can compare populations with those of a healthy stream.

A game based on population dynamics in a river is used to demonstrate the effects of pollution and to teach experimental planning and analysis of data. The game is based on the principle that species diversity in an ecosystem decreases when stress is applied. The game starts by asking how many organisms to identify. This represents someone looking through a microscope to key out organisms until the total number equals the player's request. A charge is assessed for each run based on the labor required. The problem is to generate sufficient data for valid conclusions about the relationships between populations and pollution at a reasonable cost. The computer algorithm is quite complicated but less so than nature's responses, thus the player has a fair test of his reasoning powers.

This game is based on experiences related by a biologist who classifies populations. The actual cost in 1970 was about 50 cents per organism identified, but the game inflates this cost to the 1980's. A preliminary version of the game was programmed in FORTRAN in 1971 at Clemson

University. Although considered unfinished, the game was assigned to students in a course in ecological modeling. Eventually, features were to be added to assisted in data analysis and to guide the student in planning computer runs. Much to the author's surprise, the students were highly enthusiastic about the game. A BASIC version was assigned to a graduate class in 1977 at R.P.I. Again the students were highly complimentary. One stated that it was one of the best learning experiences he ever had because the assignment was open-ended. He had to: 1. decide how much to spend on generating computer data, 2. develop a research plan, 3. interpret complicated interactions, and 4. report and justify his project. Typical computer output is shown in Fig. 6-3 and Fig. 6-4 is a listing of the program.

As natural populations are stressed, the organisms that were well acclimated and in high numbers may now encounter unsuitable conditions. If the stress is high temperature, low dissolved oxygen, extreme pH, or a toxic chemical, there may be few organisms that can thrive. These increase in numbers as others decline. Commonly, fewer species exist in a stressed environment. The game reflects these population shifts, but the player has no frame of reference for predicting the time scale or the amount of pollution that results in stress. Furthermore, randomness in the data makes it difficult to quantitify changes in the populations, so relatively large numbers of organisms must be identified but at a significant cost.

It is interesting to observe differences in student reports. The engineers tend to apply statistics or to formulate mathematical relationships while the biologists use ecological terminology and see qualitative changes. Both tend to be surprised at how expensive a simple research program can be.

MODELING OF LAKES

A multitude of lake models is found in the literature, and they range from considering the lake as a perfectly mixed lumped-parameter system to highly complicated stratified, multi-species, interacting systems with complex chemistry. The first step in modeling a lake is to decide the goals. Sufficient detail must be incorporated into the model to account for the effects of interest, but too much detail may make the model unworkable. By analogy, consider modeling the world. A perfect model would have terms for every community, every person, every animal, and all the physical, chemical, and biological systems. Of course, such a model is impossible. Consider further the problem of modeling world economics. Here the model might be useful if

Figure 6-3

TYPICAL OUTPUT FOR POPULATION DYNAMICS GAME

POPULATION DYNAMICS GAME

GAME NUMBER   5
NUMBER OF ORGANISMS TO BE KEYED OUT = 400

INITIAL DISTRIBUTION

TDH(L#B'"FS"'$DHS)SI%DMQJ"""ASQCC(AQD""E%!DP)CS%JHP"FTJ%#ALMM
D)F&C)DPBHG%HP%P)QGLHRCDHCHCM%BSMIF)SHCSANCP$CAQCH((LMH($%%CC
SRRQCF"RSA"FFDPMBDRFS#%HCPHRS%"#H)(PHR)M#%BCM"%RS)Q)H"P'#(M!!
C"HT)HS#M'(!SMFDH!%DR"RCRFANPSTM!HB"(F$EHP"B"$DI"SHRNC)DPDNEE
%'PT(BDCSSC%HBRRBC'FKTR"B!LMQF#(#

!	9	A	15	K	2
"	29	B	14	L	7
#	18	C	27	M	20
$	9	D	23	N	4
&	1	F	15	P	23
'	6	G	4	Q	12
(	14	H	40	R	22
)	18	I	3	S	24
*	1	J	7	T	7

COST = $ 200

NUMBER OF ORGANISMS TO KEY OUT = 400
AMOUNT OF POLLUTION = 20
DAYS UNTIL FIRST SAMPLE TAKEN = 20

R#GMDST%M)RD$$Q(R%Q$ESPTM"QDHCA"RTS%B%RRT&EIT"#TQ%SQ%QM#Q%DQC
D!Q"QNELHSTR%(MM#RR*%QQ#)Q!Q%I&!$!I&)STQR#!QAB$M#FMRTQ#QB&SI)
MG$LRSR$S$KD#TD$CJRDS%MSR%SQ#CHDNABT"&##!$%D#D&#M%M$#)RP)LSBE
QD%AFERQ$SSILDSEDS%ESDCN#BA!L$SRRLQ%DP)%RSRSDRETB)QTB%SE"DQDE
Q$SS"MSKM"STTGB(CEB"%S$Q)$L!CSMSLB"E)MEH$#)J#RDJI%#N"SIHMQPRF
E))#MB#F!DRQPBS#BQP"QGS)SSSD$MSTQN#NFG"RHQISSD%!QQ%S$MQMT#Q#‡
NQERR#)QMR$HCBMSTRB)(QS%DDTNHRBQ

!	10	A	5	K	2
"	13	B	17	L	8
#	27	C	7	M	23
$	22	D	26	N	8
&	6	F	4	P	7
'	0	G	5	Q	40
(	5	H	8	R	32
)	15	I	8	S	42
*	1	J	3	T	19

COST = $ 600

(Continued)

AMOUNT OF POLLUTION = 400
DAYS UNTIL FIRST SAMPLE TAKEN = 50

TSSNQIETEDNSNQC#TT##EQS%#E#SSSET#QTEEQETQTSTQTN$ESNQTQNQNNSQQ
EQ#NQCQESQNDNTENSTN#Q#TSENTDSNSEN#NEQSN##EEQRQN#TTTQQTQ!QTNQQ
ETSQSQNET#EETQET##S#SQTTQQNQEQNN#QN##EQQNQE#QTTSNQQIDSQNIDEE
T#NNENSNTNGE#$NN#ENTTTET#DQTQTQQQSQQET#TTQSET$SQQTSEQ#QN##STT
SSQNSNSTN#NTSNEETTS#ENTDNNT#TSETETTENEQQN#QE##T#B#ES#QTENDSSS
QQEBNEQQD#STEENEQT#E#NNQST#TEEQNDS#TTETQE*STNTESQQ#SN#TTNT#RR
QEQSSEEQQTNSMINEQTQETBDQ##T$QE#STQ

!	1	A	0	K	0
"	0	B	3	L	0
#	51	C	2	M	1
$	4	D	11	N	60
&	0	F	0	P	0
'	0	G	1	Q	76
(	0	H	0	R	2
)	0	I	4	S	49
*	1	J	0	T	72

COST = $ 1200
AMOUNT OF POLLUTION =-9

Figure 6-4

LISTING OF POPULATIONS DYNAMICS GAME

```
10 REM STREAM GAME
20 DIM A$(30),IN(30),BI(30),S(30)
30 PRINT:PRINT"     POPULATION DYNAMICS GAME"
40 REM GAME NUMBER PROVIDES SEED TO RANDOM NUMBER GENERATOR
50 INPUT "GAME NUMBER";X:PRINT
60 X=-X
70 T=0
80 FOR I =1 TO 30
90 S(I)=0
100 X=RND(X)
110 IN(I)=X
120 T=T+X
130 BI(I)=T
140 NEXT
150 IN(0)=RND(X)
160 FOR I=1TO30
170 BI(I)=BI(I)/T
180 NEXT
190 INPUT "TYPE NO. OF ORGANISMS TO BE KEYED OUT ";N
200 PRINT:PRINT" INITIAL   DISTRIBUTION":PRINT
210 GOSUB 410
220 PRINT:GOSUB 610
230 PRINT:PRINT
240 INPUT"NUMBER OF ORGANISMS";N
250 PRINT:INPUT"AMT. OF POLN.";PO:PRINT
260 INPUT"DAYS UNTIL SAMPLE";TM
270 IF PO=<0 THEN 370
280 T=0
290 FOR I=1TO30
300 BI(I)=IN(I)*EXP(SQR(PO)*IN(I-1)*TM/(TM+20))
310 T=T+BI(I)
320 BI(I)=T
330 NEXT
340 FOR I=1TO30
350 BI(I)=BI(I)/T
360 NEXT
370 PRINT:GOSUB 410
380 PRINT:GOSUB 610
390 GOTO 250
400 STOP
```

Continued

```
410 REM SUBROUTINE FOR GEN. POP
420 FOR I=1TO30
430 S(I)=0
440 NEXT I
450 FOR J=1TON
460 X=RND(X)
470 I=0
480 I=I+1
490 IF I=31 THEN 460
500 IF BI(I)<X THEN480
510 S(I)=S(I)+1
520 L=L+1:IF L>60 THEN570
530 IF I<11 THEN K=I+32
540 IF I>10 THEN K=I+54
550 PRINT CHR$(K);
560 GOTO 580
570 L=0:PRINT CHR$(K)
580 NEXT
590 PRINT
600 RETURN
610 FOR I=1TO10
620 PRINT CHR$(I+32);S(I),CHR$(I+64);S(I+10),CHR$(I+74);S(I+2
630 NEXT I
640 L=0:CO=.5*N+5*TM+40
650 PRINT:PRINT"COST = $";.5*N+20*TM
660 RETURN
```

many of the smaller nations were omitted. On the other hand, larger states or cities might need to be assigned extra terms or equations. It requires great judgement for deciding what to include and what to omit, and prior experience or much trial and error during the attempt at simulation may guide the modeler. A model may be judged acceptable through the process of verification. This means that data other than those used to develop the model are tested to see how well the computer results fit an actual system.

A hydraulic model of a lake is not simple because there is temperature stratification that changes during the year, rainfall that is unpredictable, and flow patterns that change. Nevertheless, with many measurements and the development of data correlations, it may be possible to construct a mathematical model that functions well. How well depends on the goals of the modeler. If the effects of various parameters can be tested quickly or if predictions that have value can be made, the model may achieve the goals of its inventor.

Models of the biology and chemistry of a lake should have reasonably good hydraulic features. In order to develop equations that can be handled without excessive computer costs, the lake may be described as one gigantic, well-mixed pot, or it may be sectioned. The question of how many sections and whether to work with horizontal or vertical slices is difficult to answer. Should variations during each day be considered, or should there be only weekly or monthly effects? Another key is lumping of species. Certainly, it is not practical to have one or more equations for each type of microorganism present. It is probably also unrealistic to consider each species of fish or of plant life. One logical grouping is: primary producers (the photosynthetic organisms called phytoplankton), grazers (the tiny zooplankton life forms that eat the photosynthetic organisms), and the larger species that feed on the zooplankton.

The interactions between terms for populations and the chemical terms in a model may be complicated. equations. For example, acid-base and buffering relationships are very important to biochemistry and microbiology, and metabolic activities generate and consume biochemicals that are acids and bases. The take up or evolution of carbon dioxide profoundly affects pH. An algal system in bright sunlight can scavenge the carbon dioxide so that the pH rises above 10. A model that considers the ionization constants of the important acids and bases and includes the kinetics of their biochemical reactions requires some programming skill as well as some sophisticated insight into microbial biochemistry. With the hydraulic features, the time considerations, oversimplified biology, and some nutrient and chemical terms, the model is already unwieldly. When biological and chemical details are amplified, the model becomes monsterous.

Careful analysis of lake models is beyond the scope of this book. Some models fit test data quite well and have predictive value. Others are merely monuments to their author's tenacity. All have some value in organizing information and in educating the inventor.

On a smaller scale, modeling fermentation has many of the problems of modeling a lake. It may be important to advance beyond the simplification of assuming perfect mixing and to handle complicated flow patterns. How much structure to incorporate for the organisms and what physiochemical factors to model are difficult decisions. The key is to state the objectives and to assess each term in each equation as to whether it gives results commesurate with the trouble for evaluating its coefficients and developing its interactions with other terms.

MODELING OF WASTE TREATMENT

There is sufficient material on the modeling of waste treatment systems or on individual steps for waste treatment for several chapters or perhaps even an entire book. The best of the models address many of the points raised with respect to modeling lakes and stream and have features about chemistry, biology, hydraulics, and engineering. Waste treatment models will not be reviewed here, but a bibliography of some of the relevant articles may be helpful to anyone who wishes to pursue this topic further.

CONCLUSION

A broad definition of biochemical engineering certainly should include environmental engineering. However, the desire to keep this book from becoming too diffuse has led to rather brief discussion of environmental systems. The models in this field are extensive, detailed, and elaborate. The general principles are much the same as those described in other chapters.

PUBLICATIONS ABOUT WASTE TREATMENT

Aarinen, R., J. Tirkkonen, A. Holme (1978). "Experiences on Instrumentation and Control of Activated Sludge Plants - a Microprocessor Application", Preprint, VII, IFAC World Congress, Helsinki.

Adams, C.E. et al., (1975) "A Kinetic Model for Design of Completely-mixed Activated Sludge Treating Variable-strength Industrial Wastewaters," Water Res. (G.B.) 9:37.

Agrawal, P., C. Lee, H.C. Lim, and D. Ramkrishna (1982). "Theoretical Investigations of Dynamic Behavior of Isothermal Continuous Stirred Tank Biological Reactors", Chem. Eng. Sci., 37(3):453-462.

Andrews, J. (1974). "Dynamic Models and Control Strategies for Wastewater Processes", Water Res., 8: 261.

Andrews, J.F. (1968). "A Dynamic Model of the Anaerobic Digestion Process", Publication of the Environmental Systems Engineering Dept., Clemson Univ., Clemson.

Andrews, J.F. (1971). "Kinetic Models of Biological Waste Treatment Processes", Biotech. & Bioeng. Symposium, 2:5-33.

Andrews, J.F. (1977). "Dynamics and Control of Wastewater Treatment Plants", In: Fundamental Research Needs for Water and Wastewater Treatment Systems (J.H. Sherrard, ed.), Proc. National Science Foundation/Assoc. of Environ. Eng. Professors Workshop, Arlington, Va., pp. 83-92.

Beck, M.B. (1979b). "Model Structure Identification from Experimental Data", In: Theoretical Systems Ecology (E. Halton, ed.), Academic Press, N.Y., pp. 259-289.

Beck, M.B. (1981). "Operational Estimation and Prediction of Nitrification Dynamics in the Activated Sludge Process", Water Research, 15(12): 1313-1330.

Beck, M.B. and P.C. Young (1976). "Systematic Identification of DO-BOD Model Structure", Proc. Am. Soc. Civ. Engrs., J. Env. Eng. Div. 102:909-927.

Bryant, J.O., and L.C. Wilcox (1972). "Real-time Simulation of the Conventional Activated Sludge Process", Proc. of the Joint Automatic Control Conference of the American Automatic Control Council, pp. 701-716.

Buhr, H.O. and J.F. Andrews (1977). "The Thermophilic Anaerobic Digestion Process", Water Res. 11: 129-143

Busby, J.B., and J.F. Andrews (1975). "Dynamic Modeling and Control Strategies for the Activated Sludge Process", J. Wat. Pollut. Control Fed., 47:1055-1080.

Butler, P.B. and J.F. Andrews (1980). "A Dynamic Model for State and Parameter Estimation of the Nitrifying Activated Sludge Process", Trans. Instrument Soc. Am., 19(3).

Canale, R.P., (1970) "An Analysis of Models Describing Predator-Prey Interaction." Biotech. & Bioeng., 12:353.

Christensen, D.R. and P.L. McCarty (1975). Multi-process Biological Treatment Model", Jour. Wat. Poln. Contr. Fed. 47: 2652-2664.

Collins, A.S., and B.E. Gilliland, (1974). "Control of Anaerobic Digestion", Jour. Environ. Engr. Div., Proc. Am. Soc. Civil Engr. 100: No. EE2, 487-506.

Clifft, R.C. and J.F. Andrews (1981). "Predicting the Dynamics of Oxygen Utilization in the Activated Sludge Process", Water Poll. Control Fed., 53(7):1219-1232.

Curds, C.R. (1971). "Computer Simulations of Microbial Population Dynamics in the Activated Sludge Process". Water Res., 5:1049-1066.

Goodman, B.L., A.J. Englande, Jr. (1974) "A Unified Model of the Activated Sludge Process," J. Water. Poll. Control Fed., 46:312 (1974).

Graef, S.P. and J.F. Andrews, (1973), "Mathematical Modeling and Control of Anaerobic Digestion", AIChE Symp. Ser. 70 No. 136, 101-131.

Graef, S.P. and Andrews, J.F. (1974). "Stability and Control of Anaerobic Digestion", Water Poll. Control Fed. J., 46(4).

Gyllenberg, Mats (1982). "Nonlinear Age-Dependent Population Dynamics in Continuously Propagated Bacterial Cultures", Mathematical Biosciences : no. 1, pp. 45-74.

Haas, C.N. (1981). "Application of Predator-prey Models to Disinfection", Water Poll. Control Fed. J., 53(3):378-386.

Hill, D.T. and R.A. Nordstedt, (1977) "Modeling Techniques and Computer Simulation of Agricultural Waste Treatment Processes," Univ. of Fla., Gainesville, paper given at 1977 Ann. Mtg. ASAE, N.C. State Univ., Raleigh, N.C., June 26-29

Holmberg, A. (1980). "Modeling of the Activated Sludge Process for Microprocessor Estimation of Oxygen Utilization and Growth Rates", Helsinki University of Technology Systems Theory Lab. report.

Holmberg, A. (1981). "Microprocessor-based Estimation of Oxygen Utilization in the Activated Sludge Wastewater Treatment Process", International J. of Systems Sci., 12(6):703-718.

Holmberg, A. (1982). "Modeling of the Activated Sludge Process for Microprocessor-based State Estimation and Control", Water Research, 16(7):1233-1246.

Jacquart, J.D., D. Lefort, and J.M. Ravel, "An Attempt to Take account of Storage in the Mathematical Analysis of Activated Sludge Behavior", In: "Advances in Water Poll. Res.", (S.H. Jenkins, ed.), Pergamon Press, New York, N.Y., 367 pp. (1973).

Jones, P.H., "A Mathematial Model for Contact Stabilization Modification of the Activated Sludge Process." In: "Advances in Water Pollution Research." (S.H. Jenkins, ed.), Pergamon Press, New York, II - 5/1 (1971).

Kleinstreuer, C. and T. Poweigha (1982). "Dynamic Simulator for Anaerobic Digestion Processes", Biotech. Bioeng., 24(9):1941-1951.

Knowles, G., A.L. Downing, M.J. Barrett (1965). "Determination of Kinetic Constants for Nitrifying Bacteria in Mixed Culture with the aid of an Electronic Computer", J. Gen. Microbiol., 38:263-278.

Ko, K.Y., B.C. McInnis, and G.C. Goodwin (1982). "Adaptive Control and Identification of the Dissolved Oxygen Process", Automatica, vol. 18, no. 6, pp. 727-730.

Lech, R.F., H.C. Lim, C.P.L. Grady, L.B. Koppel (1978). "Automatic Control of the Activated Sludge Process - I. Development of a Simplified Dynamic Model", Water Res., 12:81-90.

Lijklema, L. (1973). "Model for Nitrification in the Activated Sludge Process", Environ. Sci. Technol., 7:428-433.

Lohani, B.N. (1981). "Modeling and Computer Simulations of Anaerobic Digestion Process by DNYAMO", Computers and Operations Research 8(1):39-48.

Marsili-Libelli, S. (1980b). "Reduced Order Modeling of the Activated Sludge Process", J. Ecol. Model., 9:15-32.

McKinney, R.E., "Mathematics of Complete-Mixing Activated Sludge." J. San. Eng. Div., Proc. ASCE, Vol. 88 (SA3):87 (1962).

Olsson, G., O. Hansson (1976). "Modeling and Identification of an Activated Sludge Process", IV - IFAC Symposium on Identification and System Parameter Estimation, Tbilisi, USSR, pp. 134-146.

Roques, H., S. Yue, S. Saipanich, B. Capdeville (1982). "Is Monod Approach Adequate for the Modelisation of Purification Processes Using Biologial Treatment?", Water Research, 16(6):839-847.

Smith, R. and J.F. Roesler, "Time Dependent Computer Model for Anaerobic Digestion," U.S. EPA, Cincinnati, OH (1976).

Sykes, R.M. (1981). "Limiting Nutrient Concept in Activated Sludge Models", Water. Poll. Control Fed., 53(7): 1213-1218.

Szetela, R.W. and T.Z. Winnicki (1981). "A Novel Method for Determining the Parameters of Microbial Kinetics. Biotech. Bioeng., 23(7): 1485-1490.

Tsuno, H.,T. Goda, and I. Somiya, "Kinetic Model of Activated Sludge Metabolism, Its Application to the Response of Qualitative Shock Load Water Res. (G.B.) 12:513 (1978).

Williams, J., "Mathematical and Physical Modeling of Anaerobic Digestion Masters thesis, R.P.I., Troy, N.Y. (1978).

Chapter 7

SELECTED TOPICS

Several teaching games and simulation exercises do not fit into the patterns of other chapters and are not sufficient in content for a chapter of their own. Sterilization of fermentation media, chromatographic separations, and mass transfer in microbial slimes have little in common, but the total effect of this chapter should be increased appreciation of using computers in biochemical engineering.

STERIL: Game for Fermenter Sterilization

Bulk sterilization of fermentation liquids almost always employs heat. The only alternative that has found much use on a large scale is passage through filters with very fine pore sizes that retain bacteria. Heat-sensitive materials are made up in concentrated solutions that are filter-sterilized and added to the rest of the medium that has been heat-sterilized. Other methods of sterilization that are used occassionally are: killing with gases such as ethylene oxide or propylene oxide, exposure to radiations from radioisotopes; shocking with electrical discharges in the liquid medium; sequential freezing, germination, and thawing carried through several cycles; and irradiation with ultraviolet light.

Usually, a fermenter and its medium are sterilized together, and all connecting lines are flushed with steam. A fermenter may be sterilzed empty and filled with sterile medium from a continuous sterilizer. As with filter sterilization, it is common practice to sterilize water and insensitive materials in the fermenter so as to minimize the volume that goes through the filter or the continuous sterilizer. It is impractical to heat and to cool a large vessel very rapidly, so there may be prolonged periods at elevated temperatures. A continuous sterilizer can bring a fluid to the killing temperature very rapidly and cool it quickly. As the activation energies for destruction of biochemicals differ markedly from the usual activation energies for killing, continuous sterilization tends to be much superior to batch sterilization in terms of preserving the nutritive value of the medium.

Typical heating and cooling curves for the sterilization cycle of a large fermenter are shown in Fig. 7-1. Lowering the pressure can cause flash cooling, but pressure may be retained so that the contents do not vaporize and foam out. There is heat damage to nutrients during both the heat up and cool down periods and a greater rate of destruction at the killing temperature. By contrast, heat up and cool down in continuous sterilization occur in a few seconds.

The mathematics of batch sterilization have been developed by considering separately the steps of heating, holding, and cooling. Chemical reaction kinetics are governed by the Arrhenius equation for the effect of temperature on rate:

$$\text{rate} = K\, e^{-\Delta E\, T} \tag{7.1}$$

where K = a coefficient
e = base of natural logarithms
ΔE = activation energy
T = absolute temperature

Over the range of interest, the energy of activation is assumed to be constant. The key to sterilization is the wide difference in the typical activation energies for the destruction of biochemicals (nutrients) and the typical activation energies for killing microorganisms or their spores. This means that temperature can be manipulated advantageously to achieve killing without overly damaging the nutrients. The equations are taken from the literature (Wang, et al, 1979, Aiba, et al 1965.) A later edition of the second reference does not have as much detail about sterilizer design. The equations are:

$$\nabla_{total} = \nabla_{heating} + \nabla_{holding} + \nabla_{cooling} \tag{7.2}$$

$$\nabla_{heating} = A \int \exp(-\Delta E/RT)\, dt \tag{7.3}$$

$$\nabla_{holding} = A \int \exp(-\Delta E/RT)\, dt \tag{7.4}$$

$$\nabla_{cooling} = A \int \exp(-\Delta E/RT)\, dt \tag{7.5}$$

where ΔE = the activation energy

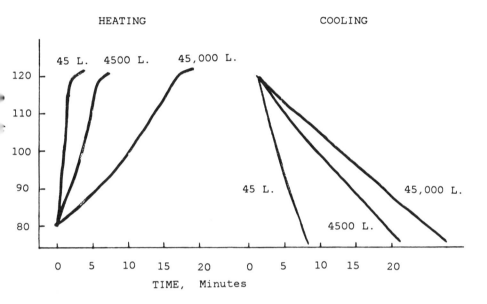

Figure 7-1

TYPICAL HEATING AND COOLING CURVES
FOR FERMENTERS OF VARIOUS SIZES

R = the universal gas constant
∇ = log of fractional reduction
A = a constant
t = time

Each integral is from zero to the time spent in that phase of sterilization.

A computer program in Appendix 6 is based on the sterilization equations with the simplifying assumption that the cooling period is the mirror image of the heating period (Bungay, 1973). It would take little additional effort to use the exact equations, but the purpose was a teaching game, and the presentation of exact results rather than a close approximation would add little. The coefficients used are not known with high accuracy anyway, thus a more rigorous set of equations has only weak justification.

It has not been possible to translate the FORTRAN version of the sterilization game to certain languages for microcomputers. The problem is the very wide range of numbers that come from the exponential terms. The game will not work when the computer language only goes to plus or minus 38 powers of ten. In fact, students must be warned that the FORTRAN version may go off range with very severe killing conditions and wrap around to give high values of organisms instead of low values. In other words, the answer may be 10^{-100} number of organisms, but the computer wraps around and reports perhaps 10^{95} organisms depending on its system for handling numbers. Some computers will just report an "Out of Range" message. If a double-precision package is available for the microcomputer, it should be possible to translate the sterilization game.

This game serves mainly for motivation. The students see that continuous sterilization can pay economic dividends and are curious about the explanations. The Arrhenius equation and the differences in activation energies have more meaning when the game has shown their importance. As the students play this game before the lectures about sterilization, the instruction are rather cryptic. They are to play the game for only 15 to 20 minutes to get a feel for the possible economic importance and to see the dramatic advantages of continuous sterilization. A typical session of play is shown in Fig. 7-2.

SETTING THE STAGE FOR THE STERILIZATION GAME

The problem for heat sterilization is to kill microorganisms without overcooking the medium. This game asks you to specify time and temperature for sterilization Organism level and concentration of heat-labile nutrients are

Figure 7-2 TYPICAL SESSION OF STERIL GAME

```
OK, R *STERIL

 TYPE STERILIZATION TEMPERATURE
121
 TYPE TIME FOR STERILIZATION
15
 ORGANISMS / FERMENTER = 0.713E-03
 % VITAMINS LEFT =  2.774
 YEARLY PROFIT = $      49562.20

 TYPE STERILIZATION TEMPERATURE
121
 TYPE TIME FOR STERILIZATION
5
 ORGANISMS / FERMENTER = 0.298E 05
         YOU LOSE, ALL THE FERMENTERS WILL BE CONTAMINATED.

 TYPE STERILIZATION TEMPERATURE
140
 TYPE TIME FOR STERILIZATION
8
 ORGANISMS / FERMENTER = 0.146E-s3     (over ran number range)
 % VITAMINS LEFT =  0.003
 YEARLY PROFIT = $   -199686.12

 TYPE STERILIZATION TEMPERATURE
140
 TYPE TIME FOR STERILIZATION
 4
 ORGANISMS / FERMENTER = 0.891E-s2011  (another overrun)
 % VITAMINS LEFT =  0.095
 YEARLY PROFIT = $   -189895.99

 TYPE STERILIZATION TEMPERATURE
 140
 TYPE TIME FOR STERILIZATION
 2
 FOR SUCH A SHORT TIME, IT IS BEST TO USE A CONTINUOUS   STERILIZER
 YOUR COSTS WILL BE FIGURED ACCORDINGLY
 ORGANISMS / FERMENTER = 0.263E-78
 % VITAMINS LEFT = 17.571
 YEARLY PROFIT = $     593555.16
```

(Continued)

```
 TYPE STERILIZATION TEMPERATURE
140
 TYPE TIME FOR STERILIZATION
1
 FOR SUCH A SHORT TIME, IT IS BEST TO USE A CONTINUOUS    STERILIZ
 YOUR COSTS WILL BE FIGURED ACCORDINGLY
 ORGANISMS / FERMENTER = 0.131E-32
 % VITAMINS LEFT = 41.918
 YEARLY PROFIT = $     909541.84

 TYPE STERILIZATION TEMPERATURE
140
 TYPE TIME FOR STERILIZATION
.5
 FOR SUCH A SHORT TIME, IT IS BEST TO USE A CONTINUOUS    STERILIZ
 YOUR COSTS WILL BE FIGURED ACCORDINGLY
 ORGANISMS / FERMENTER = 0.922E-10
 % VITAMINS LEFT = 64.744
 YEARLY PROFIT = $    1030637.76

 TYPE STERILIZATION TEMPERATURE
140
 TYPE TIME FOR STERILIZATION
.2
 FOR SUCH A SHORT TIME, IT IS BEST TO USE A CONTINUOUS    STERILI
 YOUR COSTS WILL BE FIGURED ACCORDINGLY
 ORGANISMS / FERMENTER = 0.472E 04
          YOU LOSE, ALL THE FERMENTERS WILL BE CONTAMINATED.

 TYPE STERILIZATION TEMPERATURE
145
 TYPE TIME FOR STERILIZATION
.25
 FOR SUCH A SHORT TIME, IT IS BEST TO USE A CONTINUOUS    STERILI
 YOUR COSTS WILL BE FIGURED ACCORDINGLY
 ORGANISMS / FERMENTER = 0.188E-18
 % VITAMINS LEFT = 72.795
 YEARLY PROFIT = $    1056685.87

 TYPE STERILIZATION TEMPERATURE
-2                      (This minus number stopped the program.)
```

reported along with the yearly profit that would occur if the entire factory were operated with these sterilization specifications. (Heat-labile substances are called vitamins for convenience.) One living organism is assumed to be enough to cause contamination. A number of less than I organism is converted to a percentage of contaminated fermenters, e.g. 0.1 organism/fermenter = 10% lost to contamination.

Instructions:
Login and load the STERIL program. Respond to the prompts for sterilization time and for holding temperature. The number of organisms remaining and the percentage of vitamins destroyed provide clues for the next trial. A reasonable starting trial is the laboratory condition for sterilizing shake flasks of media - 121 °C for 30 minutes. Remember that the time at damaging temperatures will be much greater for a large fermenter because of the heat up and cool down periods. In about 25 minutes of play, you will have zeroed in on a good, profitable condition. It is a waste of time to continue playing in order to increase your profit by a few cents per year.

CHROMO, A Game for Teaching Biochemical Processing

The CHROMO game is based on actual industrial problems of scaling columns from the laboratory to useful sizes (Bungay, 1978). Biochemicals such as enzymes are commonly purified by processes in which gel chromatography is a key step. A run is started by placing concentrate at the top layer of adsorbent in the column and then flushing with a solvent (eluant) that gives differential migration rates of materials through the column to its exit. The effluent is analyzed and collected in a series of containers. Fractions of high purity are combined and sent to a final step such as freeze drying. Low purity fractions may be discarded, but in the game as in real life, the fractions of intermediate purity from chromatography are valuable and are pooled for another chromatographic step.

The concentration of each material measured as it exits from the column peaks and tails off. The peaks of two materials may be so closely spaced that they coincide, or the tail of one may overlap with the rise of another. With light loading of small columns, resolution of peaks is often excellent so that very high purity is obtained. Heavier loading produces longer tails so that one material contaminates another. The dispersion of flow in large columns can also lead to severe cross contamination of materials. Loading relates to capacity of the adsorbent, some of which accept relatively heavy loading. Resolving power also depends on the nature of the adsorbent, thus a good one must be selected.

For production-scale runs, it is essential to use larger columns to achieve reasonable yields at affordable labor costs. There must be trade offs of resolving power, yields, and costs, and considerable judgement and art are required for designing a practial purification process. The player must specify: parameters of the initial purification step of ammonium sulfate precipitation, column size, column loading, and selection of one of six adsorbents with differing capacities and resolving abilities. There is a range of prices for the adsorbents, but performance does not relate to price. There can be a significant saving in selecting an adsorbent that is less expensive if its performance comes close to that of a better, more expensive adsorbent. The chromatograms are sharper if the feed stream is more pure in the desired material, but the prior purification stage becomes more expensive.

Playing chromo requires some routine screening of resins for adsorption. As in the real world, there are no manufacturer's specifications to tell how the resins function with a particular mixture. This screening should be done with relatively light loadings in order to get unsmeared chromatograms which can be interpreted. The player then scales up using a good resin and trades off loading and column diameter against resolution to get good yields. Column chromatography can function with quite impure feed streams, but resolution may suffer.

SETTING THE STAGE FOR THE CHROMO GAME

You perform research and development for a small company selling purified enzymes, and your duties include translating laboratory results to the production unit. A common step in many schemes for enzyme purification is precipitation with ammonium sulfate. Different proteins become insoluble as ammonium sulfate is added, and although the degree of purification is not great, this early step achieves some concentration so that much smaller volumes can be handled in subsequent steps.

A most useful method for purifying enzymes is gel chromatography in a packed column. It is logical for enzymes that are produced frequently to use affinity chromatography with an adsorbent that is highly specific for the desired enzyme. However, most of your products are enzymes for which sales are insufficient to justify research projects to find effective affinity agents for each one. Instead, general adsorbents are employed with which proteins move through the column at different rates. There are many adsorbents from which to choose, and the capacities and resolving powers differ for various enzymes. Note that these adsorbents are quite expensive, but there is no relation between price and performance.

One of your products is subliminal oleoaginase, a rare enzyme sold in milligram quantities for esoteric research. Someone made a supply a few years ago in one day using a very small laboratory column. The sales manager has just announced a rush order for 50 g of this enzyme from a pharmaceutical company, probably for medical testing. The implications for future sales and for developing an optimum process will not concern you yet; you are asked to work with the production group to fill this order fast. Any enzyme from test runs can be combined and sold with the main batches.

There are 6 adsorbents which are likely prospects for achieving purification of this enzyme. If you are lucky and can select a good one, there is no point in testing them all. There are no useful specifications on them because no one has ever evaluated properties with this enzyme.

You may select from your batches of partly purified enzyme. Chromatography fractions that do not meet the purity specification are saved as new batches and can be reworked by another chromatography step. A more pure feed stream means better operation of the chromatographic step, but with increased cost because of the previous losses.

The chromatograms show two other major proteinaceous impurities from which your enzyme must be resolved. Fractions of column effluent of over 94 per cent purity (the minimum specification for salable enzyme) are pooled and saved. Increased loading smears the chromatograms. Uneven flow patterns in large columns also destroy resolution. For this reason, very large columns which would allow column chromatography to be a major unit operation for the chemical process industries have never been successful. In order to save time and money, you want to use larger column diameters and moderate loadings.

DESCRIPTION OF THE PROGRAM

Functions have been selected arbitrarily to approximate the usual results with real systems. This is fairly standard in most of the author's games. A function that moves in the right direction as conditions change is quite satisfactory, and it adds little or nothing to a game to have an exact relationship. Solubilities of various proteins as functions of ammonium sulfate concentration were formulated in simple expressions and used to construct the plots shown in Figure 7-3. Note that the yield of desired enzyme is zero until a certain concentration of ammonium sulfate is reached, and

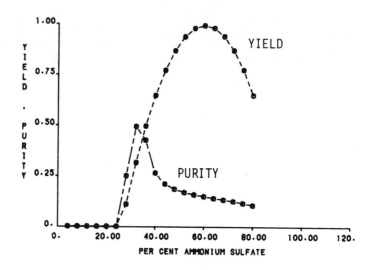

Figure 7-3

AMMONIUM SULFATE PRECIPITATION FOR CHROMO GAME

there is a peak followed by a decline in yield if too much is added. The student gains the concept that striving for high yield by adding excess ammonium sulfate causes other proteins to precipitate and contaminate the enzyme.

Simple bookkeeping routines keep track of the batches of crude enzyme prepared by ammonium sulfate precipitation. Batches may be combined, and additional batches are created automatically from side fractions in the next purification step of column chromatography.

Typical interaction with the computer is shown in Figure 7-4, and Figure 7-5 has a chromatogram from the computer. Peaks come from the equation for normal distribution. Dispersion around the center line is a function of loading of protein on the column and of purity of the enzyme. The peaks also spread causing overlap and less purification of the enzyme as column diameter increases.

There are bookkeeping routines for cost of each step and for overall expenditures. Yield of total enzyme is divided by total cost to get unit cost, and batches that are not yet sent to the chromatographic step still count in the cost calculation. Students are evaluated on economics and on the logic of their approach to the problem. Having the lowest cost for making the required amount of enzyme insures an A on the assignment, but logical experimentation may also be awarded a good grade.

This exercise gives some appreciation of the very real problem of doing as good a job as possible within time and economic restraints. Students are asked to turn in a summary of computer runs plus an explanation of their designs of each run. Results from a run made several years ago are shown to provide a starting point.

Teaching goals are: imparting a flavor of real-world biochemical engineering; illustrating operations with salt precipitation of proteins; developing appreciation of the interacting complexities when scaling up chromatographic separations; and emphasizing how economics guide industrial process development. This game accomplishes these objectives in 30 to 90 minutes of interactive play. The session is over when 50 g of product has been obtained. There is sufficient challenge to hold interest, and the concepts to be learned become fairly obvious to the player. While some other games need the instructor handy for guidance, CHROMO seems self-contained and fully self-explanatory.

Figure 7.4 TYPICAL SESSION OF CHROMO

CHROMATOGRAPHY SCALE-UP GAME

YOU MUST ACCUMULATE CRUDE ENZYME FROM AMMONIUM SULFATE STEPS
LIVER COSTS $2.48/KG UP TO 100, THEN $2.14
LABOR COST DEPENDS ON BATCH SIZE
ONE PERSON FOR <100KG
TWO PERSONS FOR <850, 3 FOR V. LARGE BATCHES.
YOU MAY ENTER CRUDE BATCHES FROM PREVIOUS RUNS.
TYPE 0 TO OMIT OR QUIT.
ENTER BATCH NO:
MG. OF ENZYME:
PURITY:
0
  KG. OF LIVER TO BE PROCESSED ?
444.
  PER CENT AMMONIUM SULFATE TO ADD ?
44.
                        COSTS
    LIVER   1056.72
    AMMONIUM SULFATE   55.29
    LABOR   902.38
    TOTAL   2014.39
    YIELD =   47987.5 MG. OR   78.26 PER CENT
    PURITY = .2134
    THIS IS ASSIGNED AS BATCH   1
    YOU NOW HAVE THE FOLLOWING BATCHES :
    BATCH 1 MG. =   47987.5 PURITY = .2134
    DO YOU WISH TO TERMINATE THIS SESSION
                                                    (YES OR NO)
NO
  TYPE A MINUS NO. TO MAKE MORE BATCHES, TYPE 0 TO
  CONTINUE TO CHROMATOGRAPHY, OR TWO NO. TO POOL BATCHES.
0
  TYPE BATCH NO. FOR CHROMATOGRAPHY
01
  ENTER MG. OF ENZYME APPLIED TO COLUMN (F FORMAT)
20000.
            THIS IS 93.731 GRAMS OF PROTEIN.
  ENTER THE DIAMETER OF THE COLUMN (F FORMAT) IN CM
20.
          THE LOADING IS   4.6865 GRAMS/LITER OF PROTEIN
          DO YOU WISH TO USE THIS LOADING (YES OR NO)
YES
ADSORBENT PRICE DOES NOT RELATE TO PERFORMANCE
BUT PRICES PER L ARE:
  ADSORBENT   PRICE
      1      $ 104.13
      2      $ 147.26
      3      $ 180.36
      4      $ 208.26
      5      $ 232.84
      6      $ 255.07
SELECT ADSORBENT FROM 1 TO 6 (I FORMAT)
5
              YIELD =   0.000 MG..   0.00 PER CENT
              OVERALL COSTS INCLUDE ALL LIVER PROCESSING

(Continued)
```
                 COSTS : $  1500.00  FOR LABOR AND ASSAYS
                          26922.25  FOR MATERIALS
                 TOTAL  $ 28422.25
          SIDE FRACTIONS OF REASONABLE PURITY ARE POOLED.
                                  TYPE ANY NUMBER TO MOVE ON.
1
  THIS IS ASSIGNED AS BATCH  2
  YOU NOW HAVE THE FOLLOWING BATCHES :
  BATCH  1  MG. =    27987.5  PURITY = .2134
  BATCH  2  MG. =    10108.4  PURITY = .4546
  DO YOU WISH TO TERMINATE THIS SESSION
                                              (YES OR NO)
NO
  TYPE A MINUS NO. TO MAKE MORE BATCHES, TYPE 0 TO
  CONTINUE TO CHROMATOGRAPHY, OR TWO NO. TO POOL BATCHES.
0
  TYPE BATCH NO. FOR CHROMATOGRAPHY
01
  ENTER MG. OF ENZYME APPLIED TO COLUMN (F FORMAT)
10000.
          THIS IS  46.865 GRAMS OF PROTEIN.
  ENTER THE DIAMETER OF THE COLUMN (F FORMAT) IN CM
20.
          THE LOADING IS    2.3433 GRAMS/LITER OF PROTEIN
          DO YOU WISH TO USE THIS LOADING (YES OR NO)
YES
ADSORBENT PRICE DOES NOT RELATE TO PERFORMANCE
BUT PRICES PER L ARE:
  ADSORBENT   PRICE
     1      $ 104.13
     2      $ 147.26
     3      $ 180.36
     4      $ 208.26
     5      $ 232.84
     6      $ 255.07
SELECT ADSORBENT FROM 1 TO 6 (I FORMAT)
6                 YIELD = 4756.2 MG.      47.563 PER CENT
                 OVERALL COSTS INCLUDE LIVER PROCESSING.
                 COSTS : $  1500.00  FOR LABOR AND ASSAYS
                          29491.85  FOR MATERIALS
                 TOTAL  $ 30991.85
                 PER MG. COST IS $     6.516
                   TOTAL MG. COLLECTED =  4756.296 AT $  12.92 PER
          SIDE FRACTIONS OF REASONABLE PURITY ARE POOLED.
                                  TYPE ANY NUMBER TO MOVE ON.
1
  THIS IS ASSIGNED AS BATCH  3
  YOU NOW HAVE THE FOLLOWING BATCHES :
  BATCH  1  MG. =    17987.5  PURITY = .2134
  BATCH  2  MG. =    10108.4  PURITY = .4546
  BATCH  3  MG. =     3589.3  PURITY = .7499
  DO YOU WISH TO TERMINATE THIS SESSION
                                              (YES OR NO)
NO
  TYPE A MINUS NO. TO MAKE MORE BATCHES, TYPE 0 TO
  CONTINUE TO CHROMATOGRAPHY, OR TWO NO. TO POOL BATCHES.
 01
 03
```
continued on next page

continued

```
   YOU NOW HAVE THE FOLLOWING BATCHES :
   BATCH  1   MG. =     21576.8  PURITY = .2422
   BATCH  2   MG. =     10108.4  PURITY = .4546
   DO YOU WISH TO TERMINATE THIS SESSION
                                                   (YES OR NO)
NO
   TYPE A MINUS NO. TO MAKE MORE BATCHES, TYPE 0 TO
   CONTINUE TO CHROMATOGRAPHY, OR TWO NO. TO POOL BATCHES.
0
   TYPE BATCH NO. FOR CHROMATOGRAPHY
01
   ENTER MG. OF ENZYME APPLIED TO COLUMN (F FORMAT)
20000.
            THIS IS  82.575 GRAMS OF PROTEIN.
   ENTER THE DIAMETER OF THE COLUMN (F FORMAT) IN CM
22.
              THE LOADING IS    3.4122 GRAMS/LITER OF PROTEIN
              DO YOU WISH TO USE THIS LOADING (YES OR NO)
YES
ADSORBENT PRICE DOES NOT RELATE TO PERFORMANCE
BUT PRICES PER L ARE:
   ADSORBENT   PRICE
       1      $ 104.13
       2      $ 147.26
       3      $ 180.36
       4      $ 208.26
       5      $ 232.84
       6      $ 255.07
SELECT ADSORBENT FROM 1 TO 6 (I FORMAT)
02
                    YIELD =   8238.0 MG.      41.190 PER CENT
                    OVERALL COSTS INCLUDE ALL LIVER PROCESSING
                    COSTS : $   1610.00   FOR LABOR AND ASSAYS
                                   0.00   FOR MATERIALS
                    TOTAL   $   1610.00
                    PER MG. COST IS $      0.195
                    TOTAL MG. COLLECTED = 12994.322 AT $     4.8!
            SIDE FRACTIONS OF REASONABLE PURITY ARE POOLED.
                                       TYPE ANY NUMBER TO MOVE
1
   THIS IS ASSIGNED AS BATCH   3
   YOU NOW HAVE THE FOLLOWING BATCHES :
   BATCH  1   MG. =      1576.8  PURITY = .2422
   BATCH  2   MG. =     10108.4  PURITY = .4546
   BATCH  3   MG. =      8729.9  PURITY = .8059
   DO YOU WISH TO TERMINATE THIS SESSION
                                                   (YES OR NO)
YES
WRITE DOWN BATCH INFO IF YOU PLAN TO RETURN.
```

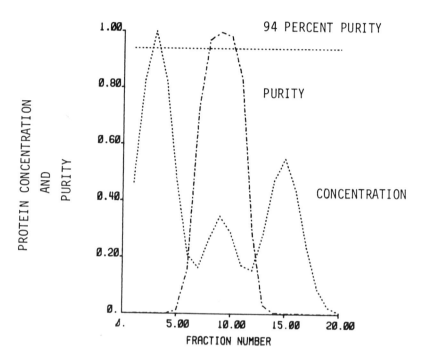

Figure 7-5

TYPICAL CHROMATOGRAM FOR CHROMO
(MIDDLE PEAK IS DESIRED ENZYME)

THIS APPEARS WITH FIG. 7.4 AFTER
SELECTION OF ADSORBENT.

The FORTRAN program for CHROMO used at the RPI Center for Interactive Computer Graphics is in Appendix 7. There is also a BASIC version that does not include all the latest features of the game because interest in BASIC faded when the graphics version was fully implemented.

MATRIX TECHNIQUES FOR MODELING MICROBIAL SLIMES

Partial differential equations can be written to model systems that have variables that are functions of both time and location. There are mathematical tricks and numerical methods for handling these equations, but often a good model can be constructed by dividing the system into a number of discrete elements. Each element is assumed to be homogeneous so that ordinary differential equations can be used. One system that has been modeled this way has nutrient medium flowing over a film of microbial slime attached to a solid surface (Bungay and Harold, 1971).

Assuming laminar conditions in the liquid, diffusion between liquid elements is negligible in the y-direction compared to flow. Only diffusion in the x-direction is considered for the elements of slime because gradients in the y-direction are much smaller. The mass balance terms are:

Flow in = concentration in previous element * area * flow rate
Flow out = concentration in this element * area * flow rate
Diffusion in = diffusivity * area * concentration difference
Diffusion out = diffusivity * area * concentration difference

Terms for mass balance equations for elements at steady state are superimposed on a crude sketch of the system in Figure 7-6. The slime is considered in vertical slices divided up into elements. The simulation used 12 horizontal sections for the flowing medium and 12 for the slime. For each element of slime, only the elements directly above and directly below deserve consideration because diffusion in the y-direction is neglected as explained previously. The elements in the slime have diffusion of oxygen and an oxygen consumption rate equal to the volume of the element times the specific uptake rate. Note that with no flow terms in the slime each element has only three terms. Viewed as a matrix, the equations have a term on the diagonal of the matrix and have terms on the adjacent diagonals. This is called a tridiagonal matrix and is common in staged operations of chemical engineering. By collecting terms, there is the equation in linear algebra:

$$A C - R + B = 0 \qquad (7.6)$$

where A   = a matrix of coefficients

$0 = $ Flow in $-$ Flow out $+$ Diffusion in $-$ Diffusion out $-$ Utilization

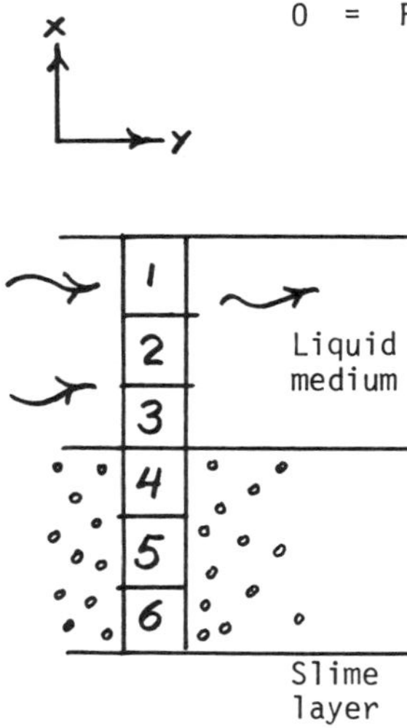

1	Liquid medium	$0 =$	$V_1 A_y C_{1,n-1} - V_1 A_y C_{1,n} + \mathcal{D}_m A_x (C_{0,n} - C_{1,n}) - \mathcal{D}_m A_x (C_{1,n} - C_{2,n})$
2		$0 =$	$V_2 A_y C_{2,n-1} - V_2 A_y C_{2,n} + \mathcal{D}_m A_x (C_{1,n} - C_{2,n}) - \mathcal{D}_m A_x (C_{2,n} - C_{3,n})$
3		$0 =$	$V_3 A_y C_{3,n-1} - V_3 A_y C_{3,n} + \mathcal{D}_m A_x (C_{2,n} - C_{3,n}) - \mathcal{D}_i A_x (C_{3,n} - C_{4,n})$
4	Slime layer	$0 =$	$\mathcal{D}_i A_x (C_{3,n} - C_{4,n}) - \mathcal{D}_s A_x (C_{4,n} - C_{5,n}) - R_{4,n}$
5		$0 =$	$\mathcal{D}_s A_x (C_{4,n} - C_{5,n}) - \mathcal{D}_s A_x (C_{5,n} - C_{6,n}) - R_{5,n}$
6		$0 =$	$\mathcal{D}_s A_x (C_{5,n} - C_{6,n}) - R_{6,n}$

where $V =$ liquid flow velocity, assume parabolic profile

$A_x, A_y =$ cross section area normal to x-direction and y-direction

$\mathcal{D}_m, \mathcal{D}_i, \mathcal{D}_s =$ diffusivity of O_2 in medium, liquid-slime interface, and slime

$C =$ oxygen concentration

Figure 7-6

STEADY-STATE EQUATIONS FOR OXYGEN TRANSFER IN MICROBIAL SLIME

C = a column of coefficients
R = a column of oxygen utilization rates
B = a column of concentrations from the previous slice

The equation used to find C is:

$$C = A^{-1}(R - B) \qquad (7.7)$$

where A^{-1} = the inverse of matrix A.

When C has been found for a slice of elements, it becomes B for the next slice. Fortunately, A does not have to be inverted over and over as the computer program advances through the model because the coefficients are very nearly constant as the oxygen concentration changes. The simulation starts with guesses for the values for the R array, finds C with Equation 7.7, and calculates the R array. There will be a difference between the assumed values and the calculated values for R, and the program iterates until R converges to nearly constant values. The procedure is repeated for slice after slice with the result that profiles of dissolved oxygen versus depth in the slime or versus distance from first contact with the slime can be drawn. The logic of the calculations is shown in Figure 7-7.

Many permutations of the oxygen transfer coefficients were tested by simulation. Results of a run that agreed fairly well with actual data are shown in Figure 7-8. Exercises such as this are valuable in confirming that the framework for interpreting experiments is sound.

There is no point in dwelling on the linear algebra approach to modeling because this book features much simpler lumped parameter systems. It is important, however, to appreciate that much more sophisticated methods are available. The model of the slime system is a rather elementary example of matrix methods.

MASS BALANCES FOR WASHING AND EXTRACTION

Washing and leaching of soluble constituents from solids are very common operations in commercial processes. The solids are contacted with the extractant, held or stirred for a time, and separated. Filtration provides effective separation, especially if the solids are squeezed to expel more of the liquid. Sedimentation is preferred over filtration because the equipment is less expensive, and multi-stage extraction requires numerous separations. As the extract often must be concentrated, the volume of extracting liquid should be

Writing the simultaneous equations in matrix algebra form:

$$
\begin{bmatrix} 0 \\ 0 \\ 0 \\ 0 \\ 0 \\ 0 \end{bmatrix} = \begin{bmatrix} -V1Ay-2DmAx & DmAx & 0 & 0 & 0 & 0 \\ DmAx & -V2Ay-DmAx & DmAx & 0 & 0 & 0 \\ 0 & DmAx & -V3Ay-DmAx-DiA & DiAx & 0 & 0 \\ 0 & 0 & DiAx & -DiAx-DsAx & -DsAx & 0 \\ 0 & 0 & 0 & DsAx & -2DsAx & DsAx \\ 0 & 0 & 0 & 0 & DsAx & -2DsAx \end{bmatrix} \begin{bmatrix} C1,n \\ C2,n \\ C3,n \\ C4,n \\ C5,n \\ C6,n \end{bmatrix}
$$

$$
- \begin{bmatrix} -DmAx\ CI \\ 0 \\ 0 \\ R4,n \\ R5,n \\ R6,n \end{bmatrix} + \begin{bmatrix} F1AyC1,n-1 \\ F2AyC2,n-1 \\ F3AyC3,n-1 \\ 0 \\ 0 \\ 0 \end{bmatrix}
$$

$$0 = AC - R + F$$
$$0 = C - A^{-1}(-R+F)$$
$$C = -A^{-1}(-R+F)$$

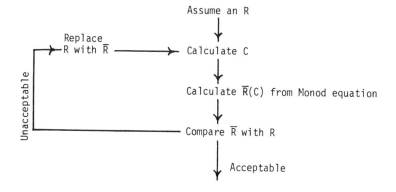

Figure 7-7
OPERATIONS FOR CALCULATING OXYGEN PROFILES

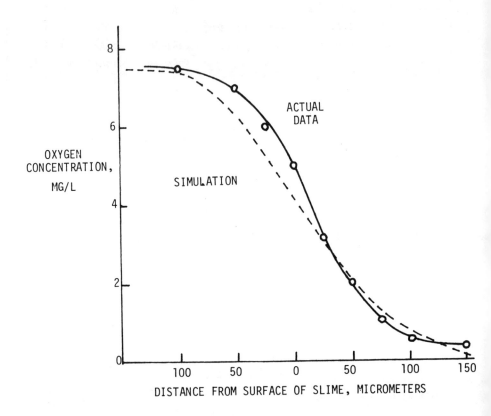

Figure 7-8

DISSOLVED OXYGEN PROFILE IN MICROBIAL SLIME SYSTEM

minimized. It is better to economize on liquid instead of paying twice - adding excess and then having additional cost for evaporation. The most efficient use of liquid is countercurrent operation where the fresh extractant contacts the spent solids and the departing liquid is brought to higher concentration by contacting the richest solids. Extraction of solids with liquids is easily handled mathematically by the techniques of undergraduate courses in staged separations of chemical engineering.

We have modified a program written by H. G. Folster of New Mexico State University for analyzing solids-liquid extraction. A few laboratory measurements of wet weight, dry weight, ratio of solids to liquid for stirring or contacting, and liquid held by the spent solids are sufficient for roughing out the parameters of a multi-stage extraction step. It is assumed that properties such as moisture content of the separated solids would not vary significantly from stage to stage. The computer program permits tinkering with the number of stages, liquid/solid ratios, and conditions for achieving specified yields.

The program for a personal computer in Figure 7-9 is in routine FORTRAN easily modified for other computers. A project is underway to convert this program to an interactive graphics teaching exercise with a menu of options including display of steps on an equilibrium diagram, presentation of a block diagram with the mass balance calculations at the input and output arrows for each block, recall of previous displays, and changing of parameters. Our use of the program to aid laboratory studies of washing and extracting exploded wood was described by Garcia-Caro, et al (1984).

MEMBRANE FOULING

Particles move across fluid streamlines and can congregate at an annulus as a suspension flows in a capillary. This is known as the 'tubular pinch effect'. Green and Belfort (1980) recognized that particle migration is complicated by fluid passage through a membrane. Particle deposition to hasten membrane fouling depends on both the pattern of flow distributed to the membrane and the flux through the membrane. Altena, et al (1983) extended the analysis of a suspension of particles flowing past a permeable membrane. Steven Fraleigh is converting the computer graphics models of these investigators to a teaching module with a menu and prompts that aid the student in testing a variety of flow rates and fluxes to visualize the deposition of particles on the membrane. A graph from a preliminary version of the teaching module is shown in Figure 7-10.

FIGURE 7-9

LEACHING PROGRAM IN NEVADA FORTRAN FOR MICROCOMPUTER

```
C     PROGRAM WRITTEN BY HARRY G. FOLSTER, NEW MEXICO STATE UNIVERSITY, AND
C     MODIFIED BY LYNN CULOTTA TO CALCULATE THE OVERALL MATERIAL BALANCE
C     AND STAGEWISE MATERIAL BALANCES FOR A LEACHING PROCESS IN WHICH SOLUTE
C     IS REMOVED FROM THE SOLID.
C
C     TWO DIFFERENT CASES MAY BE SOLVED
C        1. THE NUMBER OF EQUILIBRIUM STAGES REQUIRED TO EFFECT THE DESIRED
C           RECOVERY.
C        2. GIVEN THE RECOVERY AND THE NUMBER OF SEPARATION STAGES AVAILBLE.
C           THE QUANTITY OF SOLVENT REQUIRED IS CALCULATED.
C     VARIABLES
C     FEED      FRESH FEED
C     SOLVT     FRESH SOLVENT FEED
C     XSOLU     MASS FRACTION SOLUTE IN FEED
C     XSOLV     MASS FRACTION SOLVENT IN FEED
C     XSOLID    MASS FRACTION SOLID IN FEED
C     YSOLU     MASS FRACTION SOLUTE IN FRESH SOLVENT
C     YSOLV     MASS FRACTION SOLVENT IN FRESH FEED
C     YEXTL     MASS FRACTION SOLUTE IN PRODUCT EXTRACT
C     FRAC      FRACTION OF SOLUTE IN FEED RECOVERED
C     RATIO     WEIGHT RATIO OF WATER TO SOLID
C     NO        NUMBER OF AVAILABLE STAGES
C     ENTER THE INPUT INFORMATION BY FREE FORMAT (ONE VARIABLE PER LINE).
C
C     FOR CASE #1:  THE NUMBER OF SEPARATION STAGES, NO, MUST BE ZERO.
C
C     FOR CASE #2:  THE ESTIMATION OF THE SOLVENT RATE (SOLVT) AND THE
C                   THE NUMBER OF AVAILABLE STAGES (NO) MUST BE SPECIFIED
C                   IN THE INPUT.
C
      DIMENSION FRRFI(20),FREXO(29),RAFFI(20),EXTRO(20),RAFFO(20)
      DIMENSION EXTRI(20),FRRFO(20),FREXI(20),DIFF(200)
    1 WRITE(1,*) 'INPUT DATA'
      READ(0,*)FEED,SOLVT,XSOLU,XSOLV,XSOLID,YSOLU,YSOLV,YEXTL,FRAC,
     1RATIO,NO
C     WRITE OUTPUT TO FLOPPY DISK FILE
      CALL LOPEN(7,'OUTPUT')
      WRITE(7,118)FEED,SOLVT,XSOLU,XSOLV,XSOLID,YSOLU,YSOLV,YEXTL,
     1FRAC,RATIO,NO
C     TOTAL MATERIAL BALANCE CALCULATIONS
      SOLID=XSOLID*FEED
      SOLUT=XSOLU*FEED
      SOLV=XSOLV*FEED
      SOLUR=FRAC*SOLUT
      SOLLO=SOLUT-SOLUR
      RAFFS=SOLID*(1.+RATIO)
      RAFF=RAFFS+SOLLO
      XRAFF=SOLLO/RAFF
      K=0
      IF(NO)20,20,511
  511 NO=NO+1
```

(Continued)

```
    9 SOLVT=2.*SOLVT
   10 EXTRL=FEED+SOLVT-RAFF
      IF(EXTRL)9,9,12
   12 CONTINUE
      WRITE(1,*)EXTRL
      YEXTL=(SOLUR+YSOLU*SOLVT)/EXTRL
      K=K+1
      SNO=SOLVT
      GO TO 30
   20 CONTINUE
      EXTRL=SOLUR/YEXTL
      SOLVT=EXTRL+RAFF-FEED
   30 N=1
      FRRFI(N-1)=XSOLU
      FREXO(N)=YEXTL
      RAFFI(N-1)=FEED
      EXTRO(N)=EXTRL
C  STAGEWISE MATERIAL BALANCE CALCULATIONS
   40 CONTINUE
      SOLFO=(SOLID*RATIO/(1.-FREXO(N)))*FREXO(N)
      RAFFO(N)=RAFFS+SOLFO
      EXTRI(N+1)=RAFFO(N)+EXTRO(N)-RAFFI(N-1)
      FRRFO(N)=SOLFO/RAFFO(N)
      SOLFI=RAFFI(N-1)*FRRFI(N-1)
      SOLXO=EXTRO(N)*FREXO(N)
      SOLXI=SOLXO+SOLFO-SOLFI
      FREXI(N+1)=SOLXI/EXTRI(N+1)
      RAFFI(N)=RAFFO(N)
      FRRFI(N)=FRRFO(N)
      EXTRO(N+1)=EXTRI(N+1)
      FREXO(N+1)=FREXI(N+1)
      N=N+1
      IF(SOLFO-SOLLO)50,50,40
   50 CONTINUE
      N=N-1
      IF(NO)90,90,60
   60 CONTINUE
      IF(K-100)61,61,89
   61 IF(N-NO)70,62,80
C  NEWTON METHOD TO CONVERGE ON SPECIFIED SOLUTE RECOVERY BY
C  ADJUSTING SOLVT
   62 DIFF(K)=SOLFO-SOLLO
      IF(ABS(DIFF(K))-0.001)90,90,63
   63 FACT=SOLVT+FEED-RAFF-SOLUR
      SN1=SNO+(DIFF(K)*FACT*FACT)/(SOLID*RATIO*SOLUR)
      SOLVT=SN1
      IF(ABS(SN1-SNO)-0.01)90,90,10
C  SOLVENT RATE ITERATIONS TO OBTAIN SPECIFIED NUMER OF SEPARATION STAGES
   70 SOLVT=0.9*SOLVT
      GO TO 10
   80 SOLVT=1.10*SOLVT
```

continued on next page

(Continued)

```
          GO TO 10
      89  WRITE(7,119)
      90  WRITE(7,120)
          J=0
          WRITE(7,130)J,FEED,XSOLU,RAFF,XRAFF,SOLVT,YSOLU,EXTRL,YEXTL
          DO 100 J=1,N
          JM1=J-1
          JP1=J+1
          WRITE(7,130)J,RAFFI(JM1),FRRFI(JM1),RAFFO(J),FRRFO(J),
         1EXTRI(JP1),FREXI(JP1),EXTRO(J),FREXO(J)
     100  CONTINUE
          WRITE(7,140)
          J=N-1
          IF(NO) 101,101,102
     101  WRITE(7,150)J,N
          GO TO 159
     102  WRITE(7,155)SOLVT,FEED
    C110  FORMAT(2F7.2,7F7.4,F7.3,7X,I3)
     118  FORMAT(2X,10HINPUT DATA,/,2F7.2,7F7.4,F7.3,7X,I3,///)
     119  FORMAT(16X,66HNUMBER OF ITERATION LOOPS EXCEEDED 100, COMPUTATIONS
         1ISCONTINUED...,//)
     120  FORMAT(20X,43HSTAGEWISE CALCULATIONS FOR LEACHING PROCESS,//,32X,1
         16HMATERIAL BALANCE,//,20X,9HRAFFINATE,26X,7HEXTRACT,/,1X,5HSTAGE,1
         2(4X,5HINPUT,3X,4HFRAC,4X,6HOUTPUT,3X,4HFRAC,),5X,5HINPUT,4X,4HFRAC,
         45X,5HOUPUT,4X,4HFRAC/,5X,4(2X,8HTONS/DAY,2X,6HSOLUTE,);/,)
     130  FORMAT(2X,I3,1X,4(2X,F7.2,F9.4)/)
     140  FORMAT(2X,50HNOTE.. STAGE 0 REPRESENTS OVERALL MATERIAL BALANCE,/)
     150  FORMAT(2X,I3,2X,2HTO,I3,32H THEORETICAL STAGES ARE REQUIRED,)
     155  FORMAT(2X,F7.2,35H TONS/DAY OF WATER ARE REQUIRED PER,F7.2,17H TONS
         1/DAY OF FEED,///)
     159  CONTINUE
          WRITE(1,*)'TYPE 0 TO QUIT OR 2 TO CONTINUE'
          READ(0,*)IDUM
          IF(IDUM.EQ.0)GO TO 160
          IF(IDUM.EQ.2)GO TO 1
          GO TO 159
     160  WRITE(7,*)'END OF CALCULATIONS.'
          CALL CLOSE(7)
          STOP
          END
```

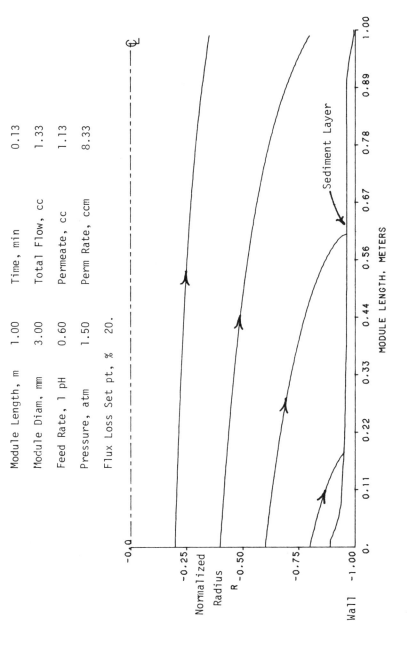

Figure 7-10

Trajectories of Particles Flowing in a Tube with Permeable Walls

This package illustrates how a rather advanced concept can be taught nicely with a computer. The analysis and its mathematics are more understandable when the student can experiment with coefficients and terms in the equations.

REFERENCES

Aiba, S., A.E. Humphrey, and N.F. Millis, (1965) "Biochemical Engineering", Univ. of Tokyo Press

Altena, F.W., G. Belfort, J. Otis, F. Fiessinger, J.M. Rovel, and J. Nicoletti (1983) "Particle Motion in a Laminar Slit Flow: A Fundamental Fouling Study" Desalination 47: 221-232

Bungay, H.R., (1973) Games for Learning", CHEM TECH, May Issue, p.290-291

Bungay, H.R., (1978) "Scale Up of Enzyme Purification" in Enzyme Engineering, Plenum Press pp.225

Bungay, H.R., (1983) "CHROMO, A Game for Teaching Biochemical Processing", in Engineering Images for the Future, ed. L.P. Grayson and J.M. Biedenbach, Proc. Ann. Conf. Am. Soc. Engr. Educ., 2: 474-494

Bungay, H.R. and D.M. Harold, (1971) "Simulation of Oxygen Transfer in Microbial Slimes", Biotechnol. Bioengr. 13: 569-579

Garcia-Caro, M., L.G. Culotta, and H.R. Bungay, (1984) "Analysis of Washing and Extraction of Steam-Exploded Wood", poster presented at Sixth Symposium on Biotechnology for Fuels and Chemicals, Gatlinburg, TN

Green, G., and G. Belfort (1980) "Fouling of Ultrafiltration Membranes: Migration and the Particle Trajectory Model" Desalination 35: 129-147

Wang, D.I.C., C.L. Cooney, A.L. Demain, P. Dunnill, A.E. Humphrey, and M.D. Lilly, (1979) "Fermentation and Enzyme Technology", Wiley-Interscience

Chapter 8

DYNAMIC ANALYSIS

There are many books on the topic of process dynamics and control, and there are several about biological systems albeit with mammallian systems as the focus. It would be absurd to try to develop the detailed dynamics of microbial systems in one chapter, but an introduction and some examples may develop an appreciation of process dynamics and show that computer graphics is a powerful tool for systems analysis.

Microbial systems are inherently dynamic. When the population grows in a simple batch culture, changes are induced in nutrients, viscosity, dissolved oxygen and carbon dioxide, waste products, pH, cell morphology, and many other factors. Natural environments exhibit drastic changes because of disturbances such as diurnal temperature and light variations, intermittent addition of nutrients, and microbial interactions about which very little is known. Continuous flow techniques in the laboratory usually give steady state populations of pure culures, but mixed cultures can be inherently oscillatory. The complex mixed culture processes for biological treatment of domestic or industrial wastes commonly undergo changes, and these can be extreme when shock loads are applied.

Engineers are primarily interested in design, operation, and control. They need data on dynamics in order to optimize their designs and to specify the correct controllers. Scientists, on the other hand, want to know why systems fluctuate, how pathways function, and how the controls of molecular biochemistry operate. Less than half of the bioengineers and bioscientists have an appreciation of dynamic analysis or any experience with it. Fortunately, there are some quite simple methods of dynamic analysis that are easy to learn and apply.

The systems engineering approach to developing a dynamic model of any process, be it electrical, mechanical, chemical, or microbiological, begins with a hypothesis of how the system works. The process is visualized as a series of interconnected steps or functional blocks. Each block corresponds to a physical or chemical step such as diffusion through a liquid, membrane transport, or an enzymatic reaction. The blocks are connected by arrows which indicate the direction of the signals (flow of information) between blocks. There may be feedback loops, alternate paths, cross connections, and other complications. It is crucial to appreciate that material flows are not represented by the connecting lines in block diagrams. Rather, information from one block is being presented to another block. Signals in a microbial system may be pH, nutrient concentration, reaction rate, oxidation-reduction potential, enzyme activity, and the like. Blocks act upon input signals to produce output signals. Figure 8-1 is a simplified block diagram, and G1 is some mathematical function.

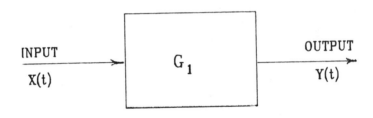

FIGURE 8-1

SAMPLE BLOCK DIAGRAM

When the interaction between system variables has been postulated in block diagram form, the functional nature of the steps represented by the blocks is evaluated. The relationship between input and output for a given block is called its transfer function, e.g. G1 in Fig. 8-1. As an example, concentration of the limiting nutrient is sensed somehow by the organisms and used as an information signal to biochemical processes that establish the specific growth rate. The mathematics relating the output signal (growth rate) to input signal (nutrient concentration) form the transfer function. Note that a combination of blocks may be required to process the nutrient information to develop the growth rate signal. If so, the transfer function for Fig. 8-1 may be quite complicated. In the future, as more biochemical details become available, a series of blocks will substitute for the single block, and each should have a relatively straight forward transfer function when the level of detail becomes that of one process per block.

A block diagram with dozens of blocks would appear to defy rational analysis, but this is not the case, Different blocks have different time responses, and those with relatively fast response have negligible effect on the timing of the entire pathway. Usually a few blocks are much slower than all the others and thus are "bottlenecks" which dominate in determining the overall response of the system. A dynamic analysis concentrates on the rate limiting steps giving little information on the many other steps, but these dominant steps are relatively much more important and can be studied in real, living situations.

Using the block diagrams or another unifying concept of how a system functions, equations are derived for time behavior. Let us represent a single input as $X(t)$, the function for the variable X with time and the output $Y(t)$, that for Y with time. Should there be several variables, there may be a number of simultaneous differential equations. Of the various techniques for solving them, one very good method for linear differential equations is Laplace transformation which is defined as a special integral (consult a math book for the definition and derivation). Transforming the functions $X(t)$ and $Y(t)$ with the Laplace operator, produces new functions $X(s)$ and $Y(s)$, i.e. $X(t)$ and $Y(t)$ are transformed from the "time domain" into the "S domain". The advantage of transforming into the S domain is that differential equations in time are transformed into algebraic equations in s. The dependent variables $X(s)$ and $Y(s)$ can then be solved for in terms of an algebraic expression in s. The time domain values are found by inversion of the Laplace transform. The use of Laplace transforms in the solution of differential equations is analogous to the use of logarithms in handling arithmetic manipulations. For example, taking a number to an uneven power is a complicated chore without logarithms and easy by taking the logaritm, multiplying by the exponent, and taking the antilogarithm.

Laplace transforms lead to a conceptual approach to systems analysis in which the s notation is meaningful in itself. Transformation to the S domain allows the expression of the functional nature of the blocks in block diagrams in terms of algebraic expressions in s. This approach is quite useful in combining blocks of a block diagram since the combination of two blocks is equivalent to their algebraic multiplication. For example in Figure 8-2, a block equivalent to the series of blocks G1, G2 and G3 would have the transfer function G1 G2 G3.

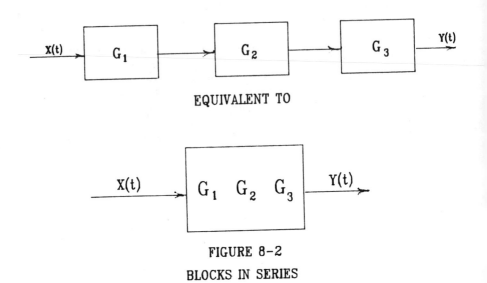

FIGURE 8-2
BLOCKS IN SERIES

TRANSIENT STATES

One of the most useful relationships, the Monod equation, for steady state or slowly changing processes becomes grossly unreliable for relating growth rate coefficients to nutrient concentrations during transient upsets. An analogy was drawn in Chapter 3 between growth and an industrial assembly line. When the rate of input of parts to the assembly line is suddenly increased, there may be some increase in production rate, but the number of available workers is soon overwhelmed to limit producion to some maximum rate. On the other hand, a sudden decrease in the feed rate of parts is quickly reflected by a decrease in production because even willing workers find little to do. Production can increase following a step up in feed rate only after a lag to recruit more workers and to create more assembly lines. Microorganisms also show a partial increase in growth rate when the feed is stepped up, and there is a time lag before the ultimate rate is reached. A step down shows little lag before the new slower rate is established.

Strange things may happen during transient states. For example, a shift from an anaerobic condition to a microaerobic system can trigger an abrupt increase in the concentration of Vitamin B12 in its fermentation. There are some reports of increased rates of production in various fermentations when they are maintained in prolonged periods of adjustment as input is intentionally varied.

FORCING FUNCTIONS

Various inputs to a system are shown as functions of time in Fig. 8-3. There are practical limitations in achieving these inputs in the real world. For example, consider the instantaneous step up of input concentration. With a biological process, sugar concentration in the reactor could be stepped up by dumping in a thick syrup, but even with intense mixing, several seconds could elapse before the concentration was uniform at the new level. Furthermore, the medium would be diluted by the addition.

Impulses or step downs are particularly troublesome when there is no brute-force means of removing the material. In the case of sugar, there are few methods for rapid removal. In theory, massive areas of dialysis membrane might remove sugar rapidly, but other soluble ingredients would be affected. Some parameters other than chemical concentrations are relatively easy to change. For example, a fairly good step down or step up in temperature is possible with good mixing and abundant heat transfer surface. It is also easy to change pH quickly with additions of strong acid or base, but salts are generated. The time scale of the system determines whether the input is close enough to ideality. Since microbial systems tend to have response times of seconds or minutes, a step that is complete in a few seconds may be perfectly acceptable. In other words, the input may be close enough to one of the ideal cases in Fig. 8-3 when the time constants are large relative to the time for the input.

The response of a system to an input function depends upon the governing equations. Consider a slender thermometer with little thermal capacitance as it is plunged into hot water. Its rate of response will be proportional to the difference between its temperature and the water temperature. The equation is :

$$dR/dt = k (T - R) \qquad (8.1)$$

where R   = reading of the theromometer
          t   = time
          k   = proportionality coefficient
          T   = water temperature

Modeling the response of this type of system was mentioned in Chapter 5 in connection with lag in growth rate adjustment. Graphs of temperature versus time for various types of forcing functions are shown in Fig. 8-4. The

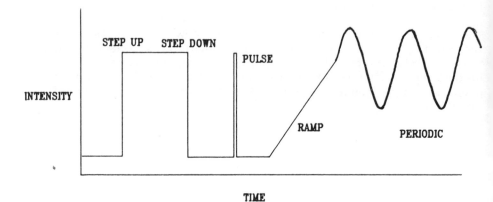

Figure 8-3
COMMON FORCING FUNCTIONS

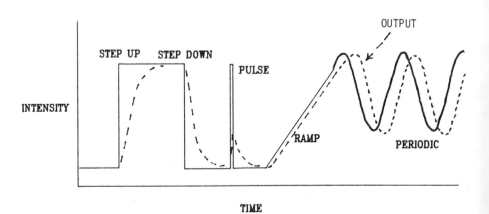

Figure 8-4
TYPICAL RESPONSES TO FORCING

thermometer is a first-order system, all of which have the transfer function

$1/(\tau s + 1)$

where τ is the time constant and s is the Laplacian operator. The time constant can usually be estimated from the properties of the system. That for a thermometer depends on the masses of glass and mercury, heat capacities, and the film coefficients and heat transfer coefficients. Some engineers prefer to estimate time constants from tabulated properties while others find it easier to plunge the thermometer into hot water while recording the temperature so that the exact time constant can be calculated. Of course, this is facietious because sometimes it is expedient to estimate time constants from system properties while other cases are better suited to direct experimentation.

Complications are introduced when the thermometer has a sheath that itself is a first-order system (because heat transfer through it is directly proportional to the temperature difference). The sheath also has the transfer function of the form

$1/(\tau s + 1)$,

but the time constant has a different value. In block diagram form, the system is shown in Fig. 8-5.

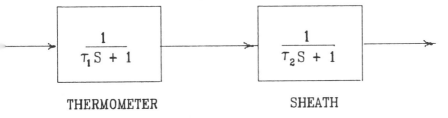

FIGURE 8-5

BLOCK DIAGRAM OF THERMOMETER WITH SHEATH

Laplace notation provides a short hand for block diagrams. The control engineer sees at a glance that Fig. 8-5 is a linear arrangement of first-order systems and has an intuitive feel for the type of response to expect.

Typical responses of first-order systems to various forcing functions are shown in Fig. 8-4. The response to a pulse is merely a response to a step interupted at a point where another response back to the initial level begins. Real systems often have unintended temporary upsets for which the responses can be used to calculate time constants. These may have the features of a pulse or a ramp. The response to a ramp seems to be excellent for calculating a time constant by simply measuring the displacement. However, getting the input to increase linearly with time can be a very difficult operational problem.

Responses to sinusoidal inputs have some very attractive mathematical properties. In fact, some of the most common terminology of control engineers is easily explained with such forcing. There may be some initial transients, but the steady-state output to a sinusoidal forcing will itself become periodic. The amplitude ratio of output over input is called the gain, and the displacement in peaks is called the phase shift. The reasons for gain and phase shift may be understood by considering a heavy truck and a light sports car moving at the same average speed as their drivers move the accellerator pedals in a sinusoidal manner. When the frequency of moving the pedal is very slow, both vehicles will track with pedal position and maintain a constant relative position. As the frequency increases, the mass of the truck comes into play so that the truck does not track as well. The truck velocity will still be sinusoidal above and below the average velocity but of less amplitude than that of the sports car. At a higher frequency, the mass of the sports car impacts on its response and the amplitude of the fluctuations in its velocity decreases. Eventually, at high frequency neither vehicle responds, and the velocities remain constant despite the rapid motion of the pedals.

Second-order systems have a more complicated response to forcing. Typical responses to step forcing are shown in Fig. 8-6. The second-order transfer function is:

$$\frac{1}{\tau^2 s^2 + 2 \zeta s + 1}$$

where τ = the time constant
ζ = the damping factor
s = the Laplacian operator

An example of a second order system is a spring. If the spring is made of lead, it will have a large damping coefficient and a response similar to Curve A. On the other hand, a slinky toy is a spring with a small damping coefficient. It may oscillate as in Curve C for many minutes.

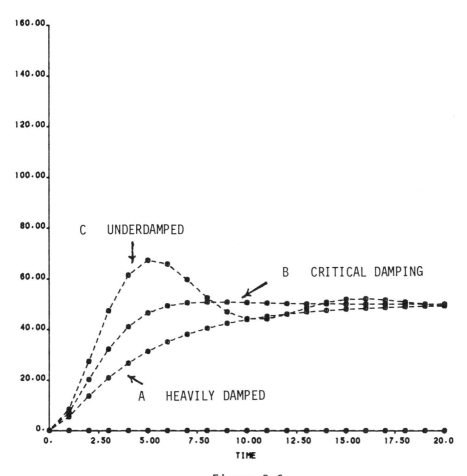

Figure 8-6

RESPONSES OF SECOND-ORDER SYSTEMS TO A STEP

Higher order systems are known, but usually combinations of first and second order transfer functions may be used. Other transfer functions of interest are those for controllers and for distance-velocity lag (also known as dead time). The simplest controller is an on-off switch that supplies corrective action when a property of a system falls above or below its desired value (set point). For example, a household thermostat is a bimetallic strip that flexes as its temperature changes. When it strikes a contact that is set at the minimum room temperature, the heating system is turned on. It takes a while before the heat circulates, so the room may drift below the desired temperature. Furthermore, heating usually overshoots the set point temperature, and conditions in the room oscillate above and below the point at which contact is made. For sensitive processes, this variation may be unacceptable, and better control modes must be employed. Proportional control applies corrective action in direct proportion to the error. Error is the difference between the actual and desired conditions. For example, if the system is far below the desired temperature, the heat would be turned to high. As the system approached the set point, heat input would be decreased. Because there must be an error to get any corrective action, proportional control has a fundamental offset - there will always be a small error even with no disturbance. Although the offset can be decreased almost to zero by employing large gain, this may cause instability. Steady-state gain is the multiplier in the transfer function when the transients have died out.

Integral control looks at the persistance of error. If error does not go to zero, the corrective action is increased. Seldom is purely integral control used; a combination is common by which proportional control dominates and integral control forces the error to disappear. Still another control mode is based on the rate of change of the error. This is termed derivative or rate control and is particularly suited to systems that can blow up. If the error is increasing rapidly, a derivative controller can institute a large amount of correction. Again, it is uncommon to use solely derivative control. An industrial controller may combine proportional, integral, and derivative control, and the relative amounts of each are adjusted for a compromise between sluggish action and stability.

Distance-velocity lag or dead time results from plug flow. For example, if the input concentration of a stream of water is varying as it enters a pipe, the same concentration relationship will be observed at the outlet of the pipe except for a shift in the time. The Laplace transform for dead time is

$$e^{-\tau s}$$

The seriousness of dead time is obvious when control is considered. Suppose that there is a delay (dead time) between the time a sample is taken and the time the analytical results are known. If the concentration must be controlled on the basis of this analytical result, the corrective action may be wrong. Figure 8-7 shows a sinusoidal input and an output delayed for some time interval. If the delay coincides with a peak in the response when the desired concentration is at a valley, the corrective action will be exactly wrong (180° out of phase) and the situation will be made worse. This is an example of an unstable control system.

Obviously, the timing of control action plays a major role in determining the stability of a system. Since the goal is to improve stability, bad control is worse than no control. Control engineers have several methods for assessing stablilty, and the terms "poles" and "zeros" are common jargon for root-locus analysis, a very powerful method for assessing system stability. Although root-locus techniques are easy enough to be part of undergraduate courses in process control, space does not permit inclusion in this book. However, poles and zero can be explained quite simply and are part of the interactive graphics method to be described later in this chapter. In some complicated transfer function such as:

$$\frac{(s + z_1)(s + z_2)}{(s + p_1)(s + p_2)}$$

if s takes on a value that cancells z1 or z2, the transfer function would be zero because a term would multiply the numerator by zero. The negatives of z1 and z2 are called the zeros of the function. By similar reasoning, should s cancel p1 or p2, division by zero in the denominator would cause the function to be infinite, so negatives of p1 and p2 are called poles of the transfer function. Plotting poles and zeros in a complex plane and observing trajectories as coefficients change is used to analyze system stability and to select appropriate values for the coefficients. The computer graphics package to be described permits analysis of poles and zeros, and the reader is encouraged to learn the rules of such analysis as given in most of the standard texts on process control. However, stability analysis is beyond the scope of this book, and only a convenient graphical rule-of-thumb will be developed.

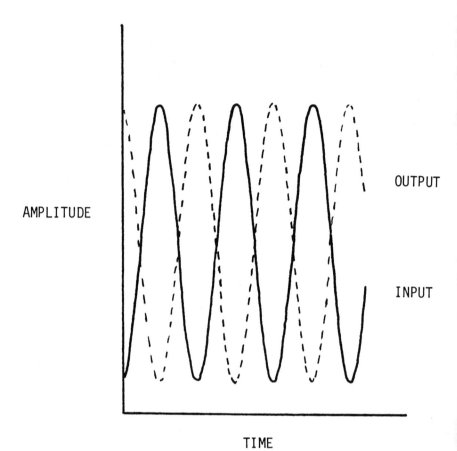

Figure 8-7

IMPOSSIBLE TO CONTROL BECAUSE OF DEAD TIME

BODE PLOTS

The frequency response of a system can be shown in plots of gain and phase shift versus frequency. Logarithm of gain and arithmetic phase shift versus logarithm of frequency are most common to compress the information into a convenient set of two graphs called the Bode diagram. Some typical Bode diagrams are shown in Fig. 8-8.

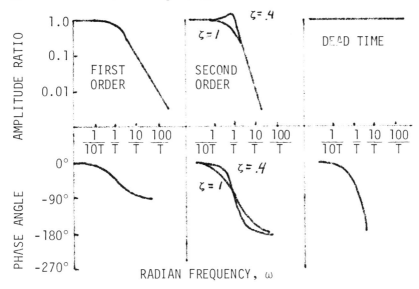

T IS THE TIME CONSTANT

Figure 8-8

BODE DIAGRAMS

At higher frequencies, the slope of the gain curve for a first-order system is -1 on the log plot, and the phase shift approaches -90°. A second-order system is characterized by a phase shift that approaches -180° and a slope of the gain curve of -2. However, there is a family of curves for second-order systems because of the damping coefficient. A low damping coefficient means that the gain can be greater than 1 in the vicinity of the frequency corresponding to the time constant. The resonance of a tuning fork at its fundamental frequency is an example. Dead time has no effect on gain but has a profound effect on the phase shift.

The mid-points in transitions of the gain curve and the phase curve relate to fundamental time constants of the system. These are the taus and the zetas of the Laplace transform notation. While there may be many steps in a process, those that are the time bottlenecks have time constants that dominate. When there are several steps in a process represented by a Bode diagram, steady-state gain of all the steps is important, but slow steps will determine the changes in gain and phase shift with frequency. Experimentally, these effects of the fast steps could not be measured. In other words, slow steps determine the response while the effects of fast steps are negligible. For example, there may be one hundred reactions in a biochemical pathway, but the one or two that are the slowest control the time behavior. A biochemist would consider this a great defect and would not use frequency response analysis to unravel the details of a pathway. On the other hand, an engineer finds it very pertinent to employ a technique that focuses on only the key steps.

Not only does the Bode diagram illustrate the time response of a system, but a general idea of stability of a system may be gleaned from the relative positions of the gain and phase curves. The gain must not be too great when the phase curve is approaching about 140° because corrective action would be badly timed and would magnify the error. Similarly, the phase shift must not be too great when the gain is high. There are rules of thumb that work fairly well for practicing engineers when specifying the settings of controllers. These are the gain margin and phase margin:

GAIN MARGIN = The amount the gain must be below zero when the phase shift is 180°.

PHASE MARGIN = The phase shift must not be greater than this when the gain is 1 (0 on a log scale)

Acceptable values for stable systems are a phase margin of -140° and a gain of -5 db (decibels) or -1.78.

In certain situations, there may be a trade off of sluggish response versus stability. Because gain margin and phase margin as defined above are somewhat conservative to insure stability, the control engineer may relax the criteria to quicken response. This is "fine tuning" of the controller and is beyond the scope of this discussion. It is important to appreciate that the transfer function of the controller is multiplied (summed with logarithms) by the transfer function of the process so that the overall Bode diagram changes. This allows adjustment of the gain and phase margins.

LINEARITY

A linear differential equation has no powers of the variables, no products of variables, and no log or trig functions of the variables. Many systems are governed by linear differential equations or the equations do not depart greatly from linearity in a narrow range of interest. When linearity prevails, the effects of several forcing funcions can be multiplied (superimposed) or the overall response is the product of the responses of individual blocks in the system. This is so important in simplifying the analysis of systems that engineers very often treat highly non-linear systems as if they were linear. The degree of error introduced by assuming linearity may be tolerable, and the systems analysis or the design of a controller may be adequate for a given task.

Bode diagram analysis is rigorous only for linear systems, and the concept of superposition allows the effects of different blocks to be handled graphically. For example, if the effects of a first-order block and a second-order block are each shown as Bode diagrams, the effect of both in series is additive on these log scales. A ruler or dividers can be used to sum the departures of each gain from one at a selected frequency and the departures of each phase shift from zero degrees.

Another handy feature of linearity is that the input forcing can be analyzed for its equivalent sine content. In theory any function can be approximated by a summation of sines or cosines. One such approach is called Fourier series analysis. This means that the input and output of a system can each be approximated by a Fourier series and compared at a range of frequencies to construct a Bode diagram. One experiment with a step or pulse input would excite all the frequencies of a system to provide data for Bode plots from which transfer functions can be derived. In contrast, a number of experiments at different frequencies of sinusoidal input would be needed to get equivalent data. However, the Fourier series approach fails with non-linear systems, and there is inadequate warning. With sinusoidal forcing, the output curve tells a story. Nonlinearities may lead to a distorted periodic output, and the amount of distortion is a clue to the amount of non-linearity.

OPEN AND CLOSED LOOP RESPONSE

When the output of a block is fed back to the input, the signal acts upon itself. The feedback can be added or subtracted. As with so many problems in process dynamics, there is a very elegant graphical method for solution. In this case, the graph called the Nichols chart relates the open loop gain and phase shift of a system to the closed loop gain and phase shift. If points are drawn over a range of frequencies, a line is defined on the Nichols chart. The trajectory of this line has meaning for control engineers and provides them a feel for the type of system.

A well-known equation for block-diagram algebra for a recycle loop is:

$$\frac{C}{R} = \frac{G}{1 + G} \tag{8.2}$$

where C = the output signal
       R = the input signal
       G = the transfer function

Complex algebra can be used to express the transfer function as:

$$G(j\omega) = A(\cos \phi + j \sin \phi) \tag{8.3}$$

where A = open loop amplitude ratio
       ϕ = open loop phase angle

Now Equation 8.2 in complex form will lead to:

$$\frac{1}{M} = \sqrt{1 + (\cos \frac{\phi}{A})^2 + (\sin \frac{\phi}{A})^2} \tag{8.4}$$

and

$$\theta = \tan^{-1} \left(\frac{\sin \frac{\phi}{A}}{1 + \cos \frac{\phi}{A}} \right) \tag{8.5}$$

where M = closed loop gain
       θ = closed loop phase shift

In lieu of using the Nichols chart, the equations may be solved directly. The calculation has been programmed, and the BASIC version is shown in Fig. 8-9.

Incorporating closed loop complications as a system is synthesized is very straight forward by modifying the transfer functions for the appropriate blocks according to Equations 8.4 and 8.5. However, the analysis of unknown systems requires some judgement. It is well known that feedback is an essential feature of many living systems. Biochemicals feed back on enzymes synthesis and activation in

```
10 REM CLOSED LOOP RESPONSE FROM OPEN LOOP
20 REM H. BUNGAY 3/78, IMPROVED BY PAT DILLON
30 DIM G(50),P(50),S(50),T(50)
40 I=1
50 INPUT"O. L. GAIN :";G(I)
60 IF G(I)<0 THEN 120
70 INPUT"O. L. PHASE :";P(I)
80 PRINT
90 K=I
100 I=I+1
110 GOTO 50
120 PRINT
130 PRINT"GAIN O.L.    PHASE O.L.    GAIN C.L.    PHASE C.L."
140 FOR J=1 TO K
150 P(J)=3.1417*P(J)/180
160 Y=COS(P(J))/G(J)
170 X=SIN(P(J))/G(J)
180 S(J)=1/SQR((1+Y)*(1+Y)+X*X)
190 T(J)=ATN(X/(1+Y))/.0174539
200 P(J)=180*P(J)/3.1417
210 IF T(J)>0 THEN T(J)=T(J)-180
220 PRINT G(J),P(J),S(J),T(J)
230 NEXT
240 PRINT
250 STOP
```

Figure 8-9   BASIC PROGRAM FOR CLOSED-LOOP RESPONSE

their metabolic pathways, and several biochemicals may act in concert for control. Frequency response data from microbial systems can be plotted as Bode diagrams, but the corner frequencies and slopes and phase shift plateaus may not agree with each other or with standard values such as 90° or 180° phase shifts. Part of the problem may arise from non-linearities and part may be blamed on closed loops. Recall that our block diagram may be provisional and oversimplified. One of its blocks may actually be a combination of blocks, some of which may have feedback. Also recall that slow steps dominate and fast steps are neglected for dynamic responses. If feedback is small or fast, it may not affect the timing of the data. When feedback is apparently an important factor, Equations 8.4 and 8.5 may be used to convert to the open loop response which may be easier to analyze.

PROGRAMS FOR DYNAMIC ANALYSIS

There are several programs available at the RPI Center for Interactive Computer Graphics that aid in dynamic analysis. One is a computer program devised by Volz, et al (1974) and modified for interactive computer graphics (Buckley, 1980, Frederick, et al, 1982). The original program, TDS (for Time Domain Solutions), has been combined with other programs for more complete graphic analysis. The user can specify the scheme of a system (this is termed the block diagram) and the time constants. Either the connections of blocks or the coefficients can be changed to determine the response and stability. There are also features for analyzing poles and zeros and for moving them on the graphs by tracking with a light pen, but an explanation is too involved for this introductory treatment.

There are several other programs available that have features comparable to the TDS package, and user preference often determines the choice. Examples in this book are from IGPALS (Interactive Graphics Program for Analysis of Linear Systems) because the graphs are a little more elegant.

A biochemical engineer could use one of these programs in a conventional manner to design controllers for pH, temperature, dissolved oxygen, and the like, but there are other, more exciting applications. Frequency response analysis is great for constructing systems, but it applies just as well for taking them apart. Actual response of a biological system can be shown as a Bode plot to determine the fundamental time constants. When an element such as a first-order system is recognized, it can be subtracted from the Bode diagram to see more clearly the effects of the remaining steps. Identifying the time dominant steps can be the key to improving response, and the effects of various experimental parameters on time constants may provide clues about the nature of the step. For example, if increasing oxygen concentration lowers a time constant, the step determining the time response is probably mass transfer through a film, and system performance can be improved by changes in agitation or aeration. On the other hand, no effect when oxygen is increased might indicate that nutrition or genetics research would be a better investment than engineering.

Figures 8-10 through 8-16 illustrate a session with IGPALS. Such exercises are expected of all seniors in chemical engineering at RPI, and graduate students in biochemical engineering are encouraged to learn one of these programs if they have no previous experience.

Possible exercises in biochemical engineering using computer graphics are:
design of a pH controller
analyzing a real Bode diagram for a chemostat
constructing a block diagram appropriate to a given microbial system.

EXAMPLE OF USING IGPALS

The instructions for using any of our programs for systems analysis are specific to the RPI Center for Interactive Computer Graphics, and a detailed presentation in this book cannot be justified. A general description follows.

After logging in, the student attaches to a special account that contains the desired program. This means that bookkeeping for computer costs is still under the original account, but other programs can be executed without moving them from one account to another. The student gets a menu of topics, one of which is electrical and systems programs. Selecting this option with the light pen produces another menu that shows IGPALS. Touching this with the light pen starts IGPALS with a grid of small squares and +'s on which to construct a block diagram. A junction for connecting signals can be located at a +, and a block can be located at a square. By touching a junction or a block and then touching a different junction or block, a connecting arrow is drawn.

A finished block diagram is shown in Fig.8-10 with a menu at the right and correction options at the bottom. Touching the menu item "BLOCK" creates Fig. 8-11 with another menu. To this point, the light pen has been ideal for interaction with the computer. Touching a block switches to Fig. 8-12 for getting numerical input to the computer. Touching the word "GAIN" allows substitution of a new value.

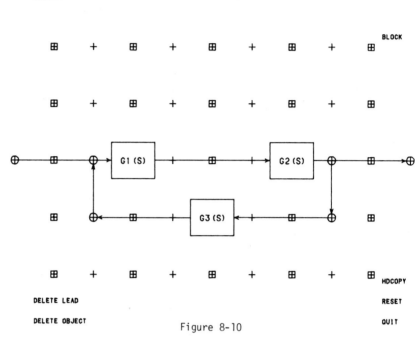

Figure 8-10

BLOCK DIAGRAM DEVELOPED WITH LIGHT PEN

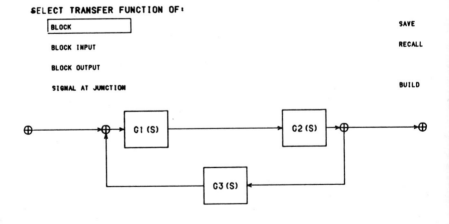

Figure 8-11

DIAGRAM READY FOR SPECIFYING TRANSFER FUNCTIONS

BLOCK 1

GAIN 1.000000

ZEROS

POLES

1. -0.500000

Block 1 has transfer function =

$$\frac{1}{s + 0.5}$$

BLOCK 2.

GAIN 100.000000

ZEROS

POLES

1. -1.000000 0.
2. -2.000000 0.

Block 2 has transfer function =

$$\frac{100}{(s+1)(s+2)}$$

BLOCK 3.

GAIN -0.500000

ZEROS

POLES

Block 3 has only gain =

-5

Figure 8-12

SCREEN OUTPUT AFTER TYPING TRANSFER FUNCTIONS

In the example, Block 1 has a first-order transfer function, $1/(\tau s+1)$, so with $\tau = 2$, the pole is -1/2. Touching "POLES" with the light pen gets the prompts 'HOW MANY POLES TO CHANGE" (student types 1), "REAL PART", (student types 0.5), and "IMAGINARY PART", (student types 0.). A composite of layouts appearing on the screen after entering specifications of the transfer functions constitutes Figure 8-12.

A menu always is at the right portion of the screen during a session with IGPALS, but these have been deleted from some of the figures to conserve space. The full menu is not part of the block diagram screen view because analysis must await specification of the parameters. The complete menu has various options. Touching 'IMPULSE' with the light pen developes a response in time as shown in Figure 8-13. Figure 8-14 results from touching 'BODE' to get plots of gain and phase shift versus frequency as in Figure 8-14. Individual blocks or combinations of blocks can be analyzed. The poles and zeros in the complex plane are shown for the overall system in Figure 8-15 that came from touching 'P Z PLOT' with the light pen. Poles and zeros can be moved to various locations to see the effects on the system responses. Two different responses are shown in Figure 8-16. The one on the left is for the system shown in the previous block diagram. The increasing amplitude of the response is characteristic of an unstable system. By returning to the BLOCK mode and changing the gain of Block 2 from 100 to 10, the stability of the system was improved. Touching 'IMPULSE' now gave the other response with decaying oscillations indicative of stability.

At any point in the interaction with the computer, a diagram can be saved for later comparisons, a hard copy can be obtained from a Versatek plotter, or the system can be changed. Any of the output plots can be rescaled by touching the scale dimensions with the light pen and then touching "LARGER" or "SMALLER". This permits regions of interest to fit the screen as the user desires. Dynamic analysis of complicated systems without a computer would be tedious and painfully slow, while analysis with a program such as IGPALS is interesting, fun, and highly effective. Many options can be surveyed in a short time. Furthermore, the student quickly develops a feel for process dynamics and can zip through a very involved analysis never realizing that this task is several orders of magnitude more sophisticated than the typical problems of just a few years ago.

For someone new to computer graphics, the use of light pens and menus to move forward and back through a complicated systems analysis must seem highly advanced.

However, this is already rather old hat to computer people. The real frontiers are with multi-dimensional plots and with color graphics, and the two-dimensional monochrome plotting of this book is considered very elementary. Engineers of the future will have tools that can handle non-linearities with ease, and the computer programs will do much of the thinking for them.

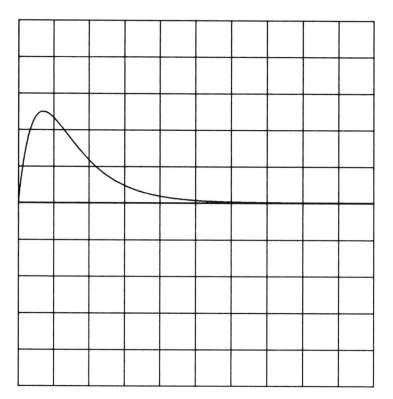

Figure 8-13

BLOCK 2 RESPONSE TO IMPULSE

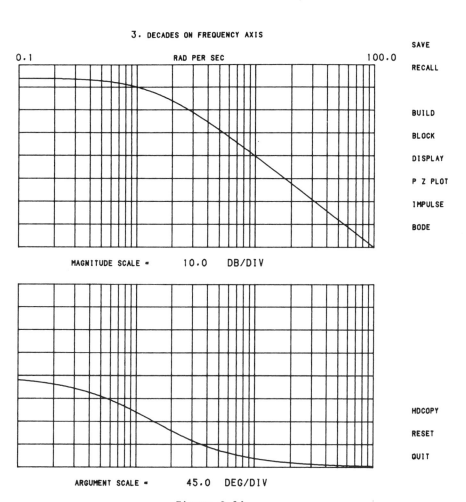

Figure 8-14

BODE PLOTS FOR BLOCK 2

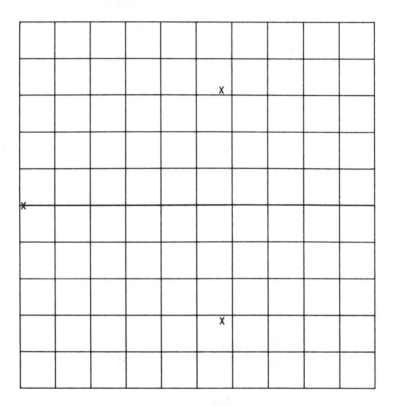

Figure 8-15
POLE-ZERO PLOT FOR CLOSED LOOP

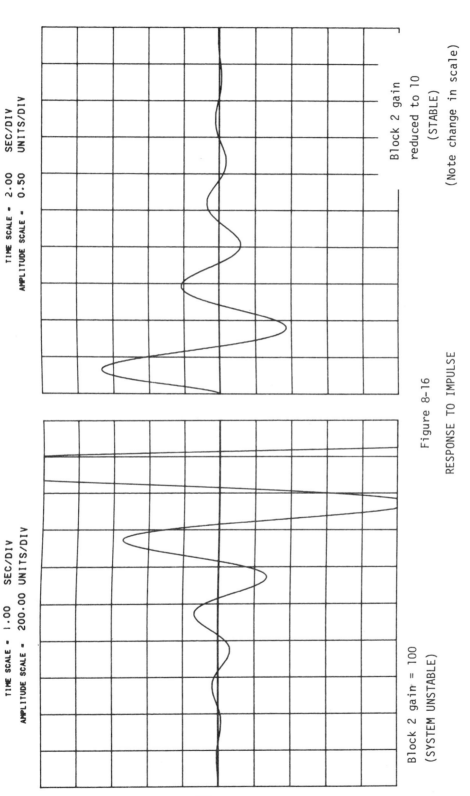

Figure 8-16
RESPONSE TO IMPULSE

REFERENCES

Buckley, P.E., (1980) "TDS, An Interactive Computer Program for Determining the Transient Response of Dynamic Systems", M.S. Project Report, R.P.I., Troy, N.Y.

Frederick, D.K., R.P. Kraft, and T. Sadeghi, (1982) "ComputerAided Control System Analysis and Design Using Interactive Computer Graphics", Control Systems Magazine, 2: 19-23

Volz, R.A., M. Dever, T.J. Johnson, and D.C. Conliff, (1974) "COINGRAD, Control Oriented Interactive Graphical Analysis and Design", IEEE Trans. Education, Vol. E-17: 3

GENERAL REFERENCES FOR CHEMICAL PROCESS CONTROL

Luyben, W.L. (1973) "Process Modeling, Simulation and Control for Chemical Engineers", McGraw-Hill.

Douglas, J.M., (1972) "Process Dynamics and Control - Vol. 1 and 2", Prentice-Hall.

Weber, T.W. (1973) "An Introduction to Process Dynamics and Control", Wiley.

Harriott, P. (1964) "Process Control", McGraw-Hill.

Tyner M., and F.P. May, (1968) "Process Engineering Control", Ronald Press Co.

Buckley, P.S. (1964) "Techniques of Process Control", Wiley.

Ray, W.H. (1980) "Advanced Process Control", McGraw-Hill.

Stephanopoulos, G., (1983) "Chemical Process Control: An Introduction to Theory and Practice", Prentice:Hall.

Hougen, J.O. "Measurements and Control Application", The Barnes & Noble Prof. Eng. Career Develop. Series.

Shinskey, F.G. (1979) "Process-Control Systems", McGraw-Hill.

Chapter 9

OTHER USES OF COMPUTERS

Most of the material so far reflects one viewpoint with examples of how games and simulation are used in one program. This chapter will touch on other uses of computers and on simulation in biochemical engineering and related disciplines. The treatment is sketchy because the intention is to provide some perspective and not to compile a treatise.

COMPUTERS FOR CHEMISTRY

A revolution is underway for organizing synthetic chemistry (Haggin, 1983). A computer console, some special programs, and a data base suffice for planning a program of chemical synthesis that takes advantage of libraries of information and predicts which paths are most likely to be successful. Some programs used by chemists are shown in Table 9-1.

Chemists dealing with synthesis of various molecules are faced with a myriad of possible routes. These computer programs list many alternatives, and some of these probably would have escaped notice if the chemist relied on memory, text books, or a literature search to compile a list in the traditional manner. Furthermore, the computer may calculate reaction energies that are an index to which reactions are spontaneous and which are unfavored. There is an underlying logic to chemical structures that should be deriveable from quantum mechanics, so we can expect computer programs to build on this logic and to become powerful tools that augment the skills of the chemist. For additional reading, see the Journal of Molecular Graphics, Butterworths Scientific, MAGSUB Ltd., Sussex, U.K.

One typical program, LHASA, for organic synthesis is described in simple terms by Long, et al, (1983). It executes on a minicomputer and has interactive graphics features. The

TABLE 9-1
COMPUTER-AIDED SYNTHESIS PROGRAMS (see Haggin, 1983)

NAME AND SOURCE	DESCRIPTION
CHIRP, Agnihotri and Motard, Washington Univ.	Uses Gibbs free energy as measure of feasibility.
LHASA, Corey, et al Harvard,	Stored reactions and connection table
SECS, Wipke, et al, Univ. of Calif., Santa Cruz,	Energy minimization routines.
REACT, Govind, U. of Cincinatti,	Reaction paths from sub-reactions.
SYNCHEM, Gelernter, et al, SUNY, Stony Brook,	Stored reactions and connection tables.
(none), Hendrickson, Brandeis,	Half-reactions and linkages.
(none), Whitlock, U. of Wisconsin	Generates synthetic routes.
(none), Bersohn, Canada,	Structures and connection tables.
CICLOPS, EROS, MATCHEM, Ugi, et al, W. Germany,	Math deductions on matrix of reactions.
SOS, Barone, et al, France	Mostly heterocyclics; connection tables.
(none), Kaufman, France,	Resembles SECS, but emphasis on organophosphorus chemistry.
MASSO, Moreau, France,	Screening system for possible routes.

goal is to suggest promising reaction pathways and to rule out those that are unreasonable. Various strategies can be selected from the program menu.

There are also interactive computer graphics programs used by chemists to examine the spatial features of structures. The images can be rotated on the display screen to see better the three-dimensional features. Promising structures for new drugs can be selected by visual examination of the representation of molecules to find the shapes induced by substituting various groups at different locations. Biochemists have also used these techniques to study enzymes. The fit of substrates to the active sites of enzymes can be explored with interactive computer graphics.

A number of games for teaching chemistry were presented at the American Chemical Society National Meeting in March, 1982. The volume number for the abstracts is 183.

COMPUTER AIDS FOR TEACHING MICROBIOLOGY

Various computer-assisted learning packages have been reviewed in the *Society for General Microbiology Quarterly*. The computer club of the Society for General Microbiology publishes a newsletter that contains descriptions of new educational programs, helpful hints, and discussion. Information is available from Dr. Barbara Evans, Department of Microbiology, University College, Newport Road, Cardiff, CF2 1TA, Great Britain. Some computer programs for microbiology are shown in Table 9-2.

TABLE 9-2
COMPUTER PROGRAMS FOR MICROBIOLOGY

Name	Description
IDEN	Player is assigned an unknown organism and given a list of tests. Objective is identification of the unknown with a minimum of tests. Bryant and Smith (1979)
PENCIL	Player varies nitrogen and sugar in a penicillin fermentation to get best yield. Calam and Russell (1973)
CCSP	Chemostat cultures. Random error is applied to data. Bazin (available from author)
COEXIST	Interactions in population dynamics. Pure and mixed culture data. Murphy (1975)
POND	Trophic level and population dynamics in a pond. Tranter and Leveridge (1973)

PREY   Prey and predator interactions. Denham (1973)

EVOLUT   Changes in differential survival rate affect composition of a population. McCormick (1975)

LINKOVER Experiments to define a genetic map. Murphy (1975)

Members of the American Society for Microbiology have formed a computer users group. For information, contact: George E. Buck, Microbiology Division, Clinical Laboratories, University of Texas Medical Branch, Galveston, TX 77550.

COMPUTER CONTROL

Computer control of laboratory equipment and of plant operations is very common. This book does not review the extensive literature about interfacing and computerized fermenters. A volume edited by Armiger (1979) shows a wide scope of computer applications to fermentation. Roels (1982) discussed aspects of kinetic models for biotechnology. Specific examples of using computers for industrial fermenters are presented by Alford (1982) and Lee and Berenbach (1982).

There are two critical areas pertaining to computer control of bioprocesses - sensors and models. Many control problems become trivial when there is a good sensor for the variable to be controlled. For example, pH electrodes are fairly rugged, reliable, and respond rapidly. Although there are some problems with breakage and some drift in the signal, it is easy and straightforward to control the pH of a bioprocess. In fact, old fashioned analog control may be a better choice than computer control unless the computer is already being used for other tasks. Other variables are not so easy to sense, and control is more difficult. When an analytical procedure must be used, there will be a delay before the results are available for determining the error and applying corrective action. As noted in Chapter 8, bad timing due to such delay can ruin control. A few sensors specific for certain products have been demonstrated in the laboratory but are not yet ready for practical applications. Some of these use enzymes in combination with pH, oxygen, or specific ion electrodes. An enzyme coating generates reaction products with the molecule being sensed, and these diffuse to the electrode.

A crucial signal missing from the arsenal of the biochemical engineer is a measure of organism concentration. Optical density is unreliable as an index of concentration or mass of organisms because interfering solids are present and because optical density changes with organism age, physiology, and properties of the medium. A major advance has been the use of computer models to relate variables that can be measured easily to those that are hard to measure. For example, pH, dissolved oxygen, exit gas concentration, rpm, and other signals available from common instruments can be fed to a computer and used to estimate a needed variable such as organism mass or product concentration. At intervals, an independent measurement of the organisms or the product can be used to correct the estimate. The algorithm depends upon an understanding of how the variables are related.

Models of the bioprocess are used for estimation, as noted above, or for prediction. Feedback control does not work well for some bioprocesses because error may appear too late for effective correction. For example, anaerobic digestion may depart from the pH at which methane production is favored, and the pH excursion defeats recovery of the methanogenic microorganisms. In other word, when the pH starts to decline, it may be too late for correction except by heroic measures. One answer to such problems is feed forward control. A good example is steering an automomobile. Feed back control would be terrible because you could generate error by how hard your vehicle was pressing against a pedestrian or a guard rail and then apply corrective action. It is much better to feed visual information to the driver's brain where a model (experience-based) predicts the position of the car and allows corrective action in advance of too much error.

Feed forward control must have a model. This model need not be perfect because there may be supplemental control. For example, there can be elements of feed back control as well. Part of the control action is based on measuring all the inputs to a process, calculating with a model what should happen, and applying corrective action in advance. At the same time, the output is sensed to determine the error, and feed back control action is taken.

MASS AND ENERGY BALANCES

Mass balance concepts have been used throughout this book to derive equations for simulation. In a larger sense, mass and energy balances are used for combinations of processes or for an entire factory. Steps can be integrated by having reactants and products flow between them, and energy sources and sinks can be coupled. This means that a hot stream from one step can be cooled by exchange with a cold stream that needs to be heated. As many steps are joined together, keeping track of mass and energy can get quite complicated. Furthermore, the bioengineer and the chemical engineer deal with processes where metabolic energy and chemical energy are important. A number of elegant computer programs have been developed for handling mass and energy in multi-step systems. These may have libraries with the thermodynamic properties of chemicals so that very elaborate calculations are handled routinely.

Kleinschrodt and Hammer (1983) provide a good example of using the programs PROCESS and HEXTRAN for designing efficient heat exchanger networks. The costs of fuel, electric power, steam, and cooling water and the desired payback time are specified to the program. Heat transfer coefficients must also be supplied, but pump horsepower and cooling duty cost are not considered. A sequence of runs can be made to find optimum flowsheets and conditions.

Discussion of computers for process engineering by Liles (1983) points out that use of chemical engineering programs increased 10-fold in ten years. The dollar benefits of using these programs and the need for further improvement were stressed. One excellent process simulation program, FLOWTRAN, was outlined by Proctor (1983). An example of computerized optimization of a process flowsheet was presented by Challand (1983).

OTHER COMPUTER PROGRAMS FOR CHEMICAL ENGINEERS

An organization exists for promoting and collecting methods for meshing computers with the teaching of chemical engineering. It is called CACHE and has the address:

> CACHE Corporation
> Room 3062 MEB
> Salt Lake City, Utah 84112
> (801)- 581-6916

They distribute a newsletter twice a year and sponsor technical meetings and short courses. An excellent series of monographs has been published about various aspects of computers and computing. Their library of computer programs includes such topics as design of heat exchangers, distillation column calculations, methods for solving both ordinary and partial differential equations, thermodynamics, and mass transfer. There are also self-teaching modules available for a variety of subjects.

The wide scope of CACHE covers just about every aspect of computer aids for design, for process control, for mathematics, for analysis, and for instruction. However, teaching games have not been exploited very much in chemical engineering.

COMPUTER TEACHING GAMES

The 1982 Yearly Index for Science Citation Index has 201 keywords under the topic of GAMES and has many cited articles. A computerized database search of computer journals and educational journals found over 1000 entries about teaching games. When the keyword "scien$" ($ means allow any letters to finish the word) was added, there were still 92 references. While many of these were for high school courses or for topics unrelated to biochemical engineering, some are quite relevant to the concepts and philosophy of computer games. Several references popped up that had little to do with games. Computerized lesson plans and interactive instruction such as PLATO, developed at the University of Illinois, may be fun but are quite distinct from game playing. It seems fair to say that there is a great deal of interest in games for education, but little has been published for the area of biotechnology. At Rutgers University, Constantinides uses computer graphics in his courses in chemical engineering and biochemical engineering and can be contacted for more information.

DRAFTING AND LETTERING

The R.P.I. Center for Interactive Computer Graphics has programs for drawing figures and for adding text. Most of the locating and moving is performed with a light pen so that progress is very rapid after becoming familiar with the techniques. Drawings in this book that were not developed with SIM4 or with a computer game were the output of a program called FONTS.DR. The program is invoked by typing R FONTS>DR to get menus for interaction with the light pen.

Lines are drawn after touching one of several different commands. To get lines parallel to the borders, HLINE or VLINE are selected. When LINE is touched with the light pen, crosshairs appear on the screen and can be positioned by dragging with the light pen. When the location is satisfactory, a function key on the terminal is struck to indicate the start of the line. The crosshairs are then dragged to the final location, and the line is drawn by again striking the function key. Using HLINE or VLINE is superior for drawing horizontals or verticals because imprecise locations of the terminating crosshairs are corrected to perfect parallels or right angles.

Menus of symbols can provide figures. They can be rotated and changed in size. Some typical symbols are shown in Fig. 9-1. The entire symbol is an entity that can be dragged about the screen with the light pen. When the diagram is finished, it is lettered by touching "TEXT" on the main menu to get a new menu. Text is entered by line or as blocks of lines. When the section of lettering is alright, depressing the function key causes it to appear in the desired type face. There are 32 faces to choose from, and Greek letters are available. As with figures, sections of lettering are dragged about the screen to position them. If a Greek letter is desired someplace in a line, a space is left, and a new line with only the Greek letter is positioned in the space. The scale factor for the lettering can be changed to get any desired size. By combining the output from the Multi-Stage Continuous Culture Game with panels of lettering produced by the FONTS.DR program, we were able to produce an elegant poster for a meeting for a cost much lower than that for employing a professional draftsman. Of course, college professors tend to overlook the cost of work by graduate students. Nevertheless, an experienced person can work very quickly with these drafting programs, and the final copy comes from a CALCOMP plotter. Various pens can produce colors or different line thicknesses.

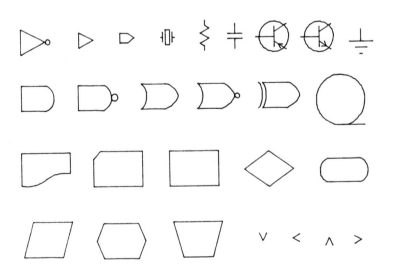

FIGURE 9-1
CONVENIENT SYMBOLS FOR DRAFTING

MISCELLANEOUS

Using computers for word processing or performing routine calculations is so common that no discussion is warranted. Languages and software determine the approaches to games and simulation. All small computers have some form of BASIC available, but until recently more powerful languages tended to cost several hundred dollars. Some still do, but PASCAL, COBOL, FORTRAN, and other fine languages are now available in very inexpensive but highly acceptable versions. The strategy of finding many customers for inexpensive software works better for some vendors than does selling to fewer customers at a high price.

There are also computer languages well-suited to special tasks such as writing programs for word processing or spread-sheet accounting. These are easier to use than assembly language which requires great attention to the step-by-step sequences of the microprocessor itself. Assembly language programs or their equivalents produced with these specialized languages tend to fit into the memory efficiently and to execute much faster than the high-level languages. Almost all languages are faster than BASIC.

Connections through telephone lines make all sorts of computer services available. Data base searching with a personal computer is fun and is effective. Most technical libraries provide similar service, but it helps to be there to suggest good keywords to the librarian. When you do the search by yourself, there is a more relaxed feeling that seems to make it easier to come up with good keywords. Another new use of computers is telephone conferencing (Turoff and Hiltz, 1977, Lerch, 1983). When groups around the country or around the world link to a large computer center, information can circulate rapidly and lively discussions ensue. Over a period of months, each participant is expected to pose questions, state opinions, submit data or examples, reply to questions, and to advance the topic of the conference. So far, the author has not had a very satisfactory experience with a telephone conference, but it is obvious that future improvements will make this a very valuable way to use computers.

There are many computer clubs that provide software free or at a nominal charge. Often a host computer is available for anyone to phone in to get messages and to see what programs are available. Contributed programs are usually accepted, and it is almost as if you were the owner of the remote computer except that there is usually a time limit to prevent abuses of the service. To get small programs, it is convenient to download through the phone lines to your own computer, but sending floppy disks through the mail is much less expensive than paying phone charges for prolonged copying. A typical service charge for a disk and its software is under $10. Popular magazines have articles containing phone numbers and addresses of computer clubs. This software ranges from not-so-good to excellent. One of the very best modem programs is in the public domain and is available from just about any computer club.

CONCLUSION

Simulation of biochemical processes can be extended to computer teaching games that generate results and determine winning or losing. Although not particularly difficult, the construction of computer games is uncommon and there is little to aid someone entering this field. Our experiences with teaching games at R.P.I. indicate considerable success. There is no doubt that some topics in biochemical engineering are nicely suited to teaching games and that sessions at the computer are invaluable for learning. The best games are interesting, challenging, and fun. They involve the student and provide an interaction not possible with other types of teaching aids. A good teaching game is effective and is usually a bargain in terms of costs and results.

When computer simulation is integrated into a research program, the scientist or engineer gains new insights. Often, it becomes a vital function because it colors the approaches to research. New ideas can be tested quickly on the computer before devoting human and financial resources to costly laboratory work. Results that contradict the computer lead to new and better analysis and eventually to better simulations. Most people that use simulation wonder how they ever got along without it.

This book has dealt with some of the principles of computer simulation by illustrating features with relatively primative programs, SIMBAS and SIM4. The reader would do well to master SIMBAS on a personal computer and then learn an up-to-date simulation package for a large computer. Even though elegant computer resources are available, the author very frequently uses SIMBAS with a personal computer in his office and switches later to better programs with the big computers if the project warrants it.

Simulation can get very complicated, and systems of simultaneous equations may be handled by advanced techniques. Special methods for partial differential equations and matrix methods for simultaneous equations were not mentioned in this book because the prime focus has been on teaching and research. The power of games and of simple simulations should be obvious at this point. Proficiency with problems of the type presented here should be a baseline for handling a high percentage of situations in biochemical engineering and for moving to more advanced techniques.

REFERENCES

Alford, J.S., (1982) "Evolution of the Fermentation Computer System at Eli Lilly and Co.", Int. Conf. on Ferm. Technol. 3/1981-08-31, Manchester, Soc. Chem. Ind. pp. 67-74

Armiger, W.B., ed. (1979) "Computer Applications in Fermentation Technology", Biotechnol. Bioengr. Symp. #9,

Bazin, E., no reference, try Queen Elizabeth College, London.

Bryant, T.N., and J.E. Smith, (1979) Jour. of Biol. Educat. 13:58-66

Calam, C.T. and D.W. Russell, (1979) Jour. Appl. Chem. and Biotechnol. 23:225-237

Challand, T.B., (1983) "Computerized Optimization of Complete Process Flowsheets", Chem. Engr. Prog. 79: 65-68

Denham, J., (1973) Computers in the Curriculum Project, Edward Arnold

Haggin, J., (1983) "Computers Shift Chemistry to More Mathematical Basis", Chem. Engr. News, May 9, p. 7-20 Vol 61 No 19

Kleinschrodt, F., and G.A. Hammer, (1983) "Exchanger Networks for Crude Units", Chem. Engr. Prog. 79: 33-38

Lee, S.E., and B.A. Berenbach, (1982) "The Use of the Minicomputer at Lederle's Fermentation Pilot Plant", Int. Conf. on Computer Applications in Ferm. Technol., Manchester, 3/1981-08-31, Soc. Chem. Ind. pp. 75-83

Lerch, I.A., (1983) "The Movable Conference", BYTE 8: 104-120

Liles, J.A., (1983) "Computer Aids for Process Engineering", Chem. Engr. Prog. 79: 43-48

Long, A.K., S.D. Rubenstein, and L.J. Joncas, (1983) "A Computer Program for Organic Synthesis", Chem. Engr. News 61: 22-30 May 9

McCormick, S., (1975) Chelsea Science Simulation Project, Edward Arnold

Murphy, P.J., (1975) Chelsea Science Simulation Project, Edward Arnold,

Proctor, S.I., (1983) "The FLOWTRAN Simulation System", Chem. Engr. Prog. 79: 49-53

Roels, J.A., (1982) "Kinetic Models in Bioengineering: Applications, Prospects and Problems", Int. Conf. on Computer Applications in Ferm. Technol., Manchester, 3/1981-08-31, Soc. Chem. Ind. pp. 37-45

Tranter, J.A., and M.E. Leveridge, (1973) Computers in the Curriculum Project, Edward Arnold

Turoff, M., and S.R. Hiltz, (1977), "Meeting Through Your Computer", IEEE Spectrum, May: 58-64

APPENDIX 1

Listing of SIM4

```
C SIMBAS IN FORTRAN WITH R.P.I. GRAPHICS CALLS
C ORIGINAL TRANSLATION BY BRUCE LOGAN
C   REVISED DECEMBER 1979 BY JAMES SALES
C  O IS X, THE OUTPUT OF THE INTEGRATION OF DX/DT
C  OIC IS THE INITIAL CONDITION OF THE OUTPUT O
C  K( ) IS A CONSTANT
C  Y IS AN ARRAY THAT CONTAINS THE VALUES OF THE OUTPUT (O)
C             WHERE Y1 WOULD CONTAIN THE OUTPUT OF O(1) AND
C             WHERE Y2 WOULD CONTAIN THE OUTPUT OF O(2),ETC.
C  ONE Y IS NEEDED FOR EACH OUTPUT   (O( )) THAT IS TO BE PLOTTED
C  N IS THE NUMBER OF DIFFERENTIAL EQUATIONS
C  TO1 IS THE INTEGRATION INTERVAL
C  TO2 IS THE TOTAL TIME
C  TO3 IS THE PRINT OR PLOT INTERVAL
C  QCALE IS THE X OR Y AXIS SCALING FACTOR
C  XLINAX IS THE X AXIS LABEL
C  YLINAX IS THE Y AXIS LABEL
C  DATLIN PLOTS THE OUTPUT AS A FUNCTION OF TIME
C  CALL HDCOPY(ICOP) COPIES THE PLOT ON THE SCREEN TO THE PLOTTER
        REAL I(100),K(100)
        DIMENSION O(100),TO(100),T6(100),TM(100),OIC(100)
        DIMENSION Y1(100),Y2(100),Y3(100),Y4(100),X(100)
C ZERO ARRAYS SO NO CRAZY PTS.
        DO 10 J=0,100
        X(J)=J
        Y1(J)=0.
        Y2(J)=0.
        Y3(J)=0.
  10    Y4(J)=0.
C ERASE SCREEN AND INIT. GRAPHICS
        CALL GRESET
C STARTING CONDITIONS
C    NUMBER OF EQUATIONS
        N=2
C    CONDITIONS OF TIME
        TO1=.1
        TO2=20.
        TO3=1.
C    INITIAL CONDITIONS
        OIC(1)=100.
        OIC(2)=0.
C INTERNAL PROGRAM STATEMENTS (DO NOT TOUCH)
        WRITE(1,25)
C ASKS FOR CONSTANTS IN D.E.'S, USED IN STATEMENTS AT 1000
```

```
      25    FORMAT('HOW MANY CONSTANTS ?')
            READ(1,26) IEQ
      26    FORMAT(I2)
       8    DO 11 J=1,IEQ
            WRITE(1,31) J
      31    FORMAT('K',I2,'=')
            READ(1,32) K(J)
      32    FORMAT(F10.5)
      11    CONTINUE
      19    T1=TO1
            T2=TO2
            T3=TO3
            DO 18 J=1,N
            O(J)=OIC(J)
      18    CONTINUE
            Y1(0)=OIC(1)
            Y2(0)=OIC(2)
            T=INT(T2/T1+0.5)
            IC=0
            T1=T2/T
            T=INT (T3/T1+0.5)
            T3=T1*T
            T=0.
            T4=0.
            IT8=1
            GO TO 1000
     100    IF ((T-T4+T1/10.).LT.0.) GO TO 125
            T4=T4+T3
            IT5=INT (T/T1+.5)
            T=FLOAT(IT5)*T1
            GO TO 2000
     115    IF ((T-T2+T1/10.).LT.0.) GO TO 125
            WRITE (1,20)
      20    FORMAT (/5X,'END OF PROGRAM')
            CALL ENTGRA
            CALL SET2D
            CALL INIT(-3.,-4.)
C END OF INTERNAL PROGRAM STATEMENTS
C PLOTTING OF OUTPUT
C     QCALE(LOWEST OUTPUT VALUE,HIGHEST VALUE, AXIS LENGTH (INCHES)
C           0=ABSCISSA,1=ORDINATE)
C     OUTPUT VALUES AND AXIS LENGTH ARE EXPRESSED IN DECIMAL FORM
            CALL QCALE(0.,T2,8.,0)
            CALL QCALE(0.,OIC(1),8.,1)
```

```
C     X OR YLINAX(LABEL IN PARENTHESES, NO. OF CHAR IN LABEL,TYPE SI
      CALL XLINAX('TIME',4,8.)
      CALL YLINAX('CONC.',5,8.)
C     DATLIN(ARRAY ON X, ON Y,NO OF PTS.,LINE TYPE, SYM., WHEN USE S
C     NEED A DATLIN FOR EACH LINE TO BE PLOTTED
C     BE CAREFUL WITH NUMBER OF POINTS IN DATLIN
      CALL DATLIN(X,Y1,21,2,1,1)
      CALL DATLIN(X,Y2,21,4,1,1)
C     INTERNAL PROGRAM STATEMENTS (DO NOT TOUCH)
      CALL EXITGR
 13   CALL TNOUA('ENTER CODE:',11)
      CALL TIDEC(IAN)
      IF(IAN) 1,2,3
 3      IF(IAN-2) 4,5,6
 6      IF(IAN-4) 7,8,9
 4      WRITE(1,33) IAN
C     CODE MINUS=QUIT,0=ERASE,1=T1,2=T2,3=T3,4=CONSTS.,5=IC'S
C     6=HARD COPY
 33     FORMAT('T',I1,'=')
        READ(1,32) TO1
        GO TO 19
 5      WRITE(1,33) IAN
        READ(1,32) TO2
        GO TO 19
 7      WRITE(1,33) IAN
        READ(1,32) TO3
        GO TO 19
 9      IF (IAN.EQ.6) GO TO 38
        DO 12 J=1,N
        WRITE(1,34) J
 34     FORMAT('IC',I2,'=')
 12     READ(1,32) OIC(J)
        GO TO 19
 1      CALL EXIT
 2      CALL GRESET
        GO TO 13
 38     CALL ENTGRA
        CALL HDCOPY(ICOP)
        CALL EXITGR
        GO TO 13
 999    CALL EXIT
 125    IF (IT8.EQ.1) GO TO 300
        IF (IT8.EQ.2) GO TO 400
        IF (IT8.EQ.3) GO TO 500
        IF (IT8.EQ.4) GO TO 600
        WRITE (1,30)
 30     FORMAT (///3X,'ERROR')
        CALL EXIT
```

```
      300 DO 320 IT5=1,N
          TO(IT5)=T1*I(IT5)
          T6(IT5)=O(IT5)
      320 O(IT5)=O(IT5)+TO(IT5)/2.
          T=T+T1/2.
          IT8=2
          GO TO 1000
      400 DO 420 IT5=1,N
          T7=T1*I(IT5)
          TO(IT5)=TO(IT5)+2.*T7
      420 O(IT5)=T6(IT5)+T7/2.
          IT8=3
          GO TO 1000
      500 DO 520 IT5=1,N
          T7=T1*I(IT5)
          TO(IT5)=TO(IT5)+2.*T7
      520 O(IT5)=T6(IT5)+T7
          IT8=4
          T=T+T1/2
          GO TO 1000
      600 DO 620 IT5=1,N
          O(IT5)=T6(IT5)+(TO(IT5)+T1*I(IT5))/6.
      620 CONTINUE
          IT8=1
C END OF INTERNAL PROGRAM STATEMENTS
C EQUATIONS AND COMPUTATIONS
C    FIRST STATEMENT MUST HAVE THE STATEMENT LABEL 1000
 1000    I(1)=-K(1)*O(1)
         I(2)=K(1)*O(1)-K(2)*O(2)
         GO TO 100
C        OUTPUT,   FILL PLOTTING ARRAYS
 2000 IC=IC+1
      X(IC)=T
C PUTTING THE OUTPUT (O( )) INTO THE Y ARRAYS
C ONE Y ARRAY NEEDED FOR EACH OUTPUT PLOTTED
      Y1(IC)=O(1)
      Y2(IC)=O(2)
C PRINTING OF THE OUTPUT
          WRITE(1,29) T,O(1),O(2)
   29     FORMAT(F3.0,2X,F10.3,2X,F10.3)
          GO TO 115
          END
C
```

APPENDIX 2

FORTRAN Listing of FERMT

```
C FERMT GAME, THIS VERSION READS A FILE OF DATA
        DIMENSION A(10)
C COMMANDS ARE FOR CP/M SYSTEMS, READ DATA FROM FLOPPY DISK FILE
C NEVADA FORTRAN
C SEE REFERENCE FOR ORIGINAL VERSION, IT HAS COMMENTS
C BUNGAY, PROCESS BIOCHEM. 6: 38 (1971)
        CALL OPEN(4,'FERMT.DAT')
        CALL OPEN(5,'FERMT.OUT')
        WRITE(5,15)
        C=0.0
        CO=0.0
1       DO 80 I=1,7
        READ (4,*,END=999) A(I)
        IF(A(I).LT.0.)GO TO 60
        IF(A(I).GT.10.) GO TO  60
80      CONTINUE
        WRITE(5,50)
        WRITE(5,30) A(1),A(2),A(3),A(4),A(5),A(6),A(7)
        C=50*A(2)+2543*A(1)+10*A(6)+15*A(4)
        C=C+18*A(5)+200*A(7)
        P3=A(4)
        S=A(2)+1.1*A(3)+0.05*A(4)+0.06*A(5)
        T=0
        G=A(3)
        X=0.02*A(1)
        A4=A(6)**4
        V=A(7)+0.05*A(4)+0.09*A(5)
        P1=P3+0.3*A(5)
        P2=A(5)+0.2*A(4)
110     G1=0.12*S/(0.4+S)
        A1=G*V/((G+0.5)*(V+0.02+V*V*300))
        D1=G1*X
        D2=3*A(6)
        IF(X.LT.D2)GO TO 90
        D1=D1*3*A(6)/X
90      X=X+D1
        T=T+1
        G=G+0.03*A4
        S=S-1.3*D1+0.02*A4
        IF(S.LT.0.05)GO TO 10
        IF(T.LT.96) GO TO 110
10      WRITE(5,120)T
```

```
              G=G+0.03*A4*(96-T)
              CO=C+100*G
              A2=1
              D3=5*A(6)
              IF(X.LT.D3) GO TO 130
              A2=4*A(6)/X
      130     Y7=(2+G)/(3*G)
              Y1=1.6*P1/(0.01+0.3*P1*P1+0.8*P1*P2)
              Y8=V/(0.003+0.1*V+10*V*V)
              Y=Y8*Y1*Y7*X*A2*(96-T)
              P8=(183*Y)-(9.1*CO)-48531
              WRITE(5,140)X,G,Y,P8
              GO TO 40
      60      WRITE(5,160)
      40      CONTINUE
              GO TO 1
              WRITE(5,150)
      15      FORMAT(25X,'   FERMENTATION GAME')
      30      FORMAT(7F10.4)
      50      FORMAT(/,'  INOCULUM    SUGAR     OIL    SOYMEAL   DIST
         1SOL     AIR     VITAMINS ')
      120     FORMAT(,'PRODUCT FORMATION BEGAN AT ',F5.2 ,'HOURS',)
      140     FORMAT(/,'TOTAL GROWTH=',F10.3,5X,'TOTAL OIL =',F10.3,
         1 5X,'YIELD =',F10.3,//,30X,,'PROFIT = $',F12.2,)
      150     FORMAT(30X,'*** END OF EXPERIMENT ***')
      160     FORMAT(20 X,'RUN CONTAMINATED!',)
      999     CALL CLOSE(4)
              CALL CLOSE(5)
              STOP
              END
```

BASIC Listing of FERMT

```
10 DIM N(71)
20 PRINT
30 PRINT TAB(19),"FERMENTATION GAME"
40 FOR J=1 TO 8
50 INPUT"INOC.    ";N(J)
60 INPUT"SUGAR    ";N(J+10)
70 INPUT"OIL      ";N(J+20)
80 INPUT"SOYMEAL  ";N(J+30)
90 INPUT"DIST. SOL";N(J+40)
100 INPUT"AIR     ";N(J+50)
110 INPUT"VITAMINS";N(J+60)
120 PRINT
130 NEXT J
132 C=0
133 CO=0
135 PRINT
140 FOR J=1 TO 8
150 PRINT"RUN INOC.   SUGAR    OIL    SOYMEAL DIST.SOL. AIR
160 PRINT J;TAB(5);N(J);TAB(13);N(J+10);TAB(21);N(J+20);TAB(2
162 PRINT N(J+30);TAB(37);N(J+40);TAB(45);N(J+50);TAB(53);N(J
170 FOR K= 0 TO 60 STEP 10
180 IF N(J+K)>0 THEN 220
190 PRINT"RUN CONTAMINATED"
210 GOTO 690
220 IF N(J+K)>10 THEN 190
230 NEXT K
240 LET C=50*N(J+10)+20*N(J)+10*N(J+50)+15*N(J+30)+18*N(J+40)
250 LET C=C+200*N(J+60)
260 LET P3=N(J+30)
270 LET S=N(J+10)+1.1*N(J+20)+.05*N(J+30)+.06*N(J+40)
280 LET T=0
285 LET G=N(J+20)
290 LET X=.02*N(J)
300 LET A4=N(J+50)'4
310 V=N(J+60)+.05*N(J+30)+.09*N(J+40)
320 LET P1=P3+.3*N(J+40)
330 LET P2=N(J+40)+.2*N(J+30)
340 LET G1=.12*S/(.4+S)
350 LET A1=G*V/((G+.5)*(V+.02+V*V*300))
360 LET D1=G1*X
370 IF X<3*N(J+50) THEN 400
380 LET D1=D1*3*N(J+50)/X
```

```
400 LET X=X+D1
410 LET T=T+1
420 LET G=G+.03*A4
430 LET S=S-1.3*D1+.02*A4
440 IF S<.05 THEN 500
450 IF T<96 THEN 340
500 PRINT"PRODUCT FORMATION BEGAN AT";T;"HOURS."
510 LET G=G+.03*A4*(96-T)
520 LET C0=C+100*G
530 LET A2=1
540 IF X<5*N(J+50) THEN 600
550 LET A2=4*N(J+50)/X
600 LET Y7=(2+G)/(3*G)
610 LET Y1=1.6*P1/(.01+.3*P1*P1+.8*P1*P2)
620 LET Y2=P2/(.014+P2*P2+.5*P1*P2)
630 LET Y8=V/(1E-03+.1*V+10*V*V)
650 LET Y=Y8*Y1*Y2*X*A2*(96-T)
660 LET P8=100*Y-C0
670 PRINT"TOTAL GROWTH =";X,"TOTAL OIL =";G
680 PRINT"YIELD =";Y,"PROFIT = $";P8
690 PRINT
700 NEXT J
720 PRINT"END OF EXPERIMENT."
730 STOP
```

APPENDIX 3

Listing of MONOD Game

```
10 REM MONOD CHEMOSTAT GAME
20 INPUT"MUMAX";MU
30 INPUT"KS    ";KS
40 INPUT"Y     ";Y
50 INPUT"SZERO";SO
60 INPUT"M.ENG";EM
70 PRINT
80 PRINT " D              PLOT OF X AND S VERSUS D"
90 D=1E-04
100 S=KS*D/(MU-D)
110 IF S>SO OR S<0 THEN S=SO
120 X=D*Y*(SO-S)/(D+EM*Y)
130 IF X<0 THEN X=0
140 IF X>Y*SO THEN X=Y*SO
150 REM SCALING FOR PLOT
160 S=S+5
170 X=X+5
180 D1=D-1E-04
190 IF X<S THEN GOTO 220
200 PRINT D1;TAB(S);"S";TAB(X);"X"
210 GOTO 230
220 PRINT D1;TAB(X);"X";TAB(S);"S"
230 D=D+.1
240 IF D<MU+.2 THEN 100
250 STOP
```

APPENDIX 4

BASIC Listing for Recycle Program

```
10 REM EARLY VERSION OF RECYCLE FERMENTER
15 X=0:Y=0:D=0:KS=0:S=0:S0=0:EM=0
100 INPUT"MUMAX";MU
110 INPUT"KS     ";KS
120 INPUT"Y      ";Y
130 INPUT"SZERO";S0
141 INPUT "CONC. FACTOR ";C
142 INPUT"FRACTION RECYCLED ";A1
145 A=1+A1-A1*C
150 D=1E-04
165 B=D*A/(MU-D*A)
167 S=KS*B
170 IF S>S0 OR S<0 THEN S=S0
180 X=Y*(S0-S)/A
190 IF X<0 THEN X=0
200 X2=A1*D*X
210 P1=INT(192-X/40):P2=INT(192-S/40)
222 Q=15+X/40
225 PRINT D,S,X,X2
260 D=D+.025
270 IF X=0 THEN CO=CO+1
275 IF CO<6 THEN GOTO 165
280 CO=0
300 GOTO 141
```

Appendix 4B

Early Version of Two Stage Fermentation

```
10 REM TWO-STAGE CHEMOSTAT, 1ST IS OLD MONOD GAME
20 X=0:Y=0:D=0:KS=0:S=0:S0=0:EM=0
30 INPUT"MUMAX";MU
40 INPUT"KS    ";KS
50 INPUT"Y     ";Y
60 INPUT"SZERO";S0
70 INPUT"M.ENG";EM
80 INPUT "RATIO OF D2 / D1 ";R
90 D=1E-04
100 PRINT"    D            X1            S1            X2
110 PRINT
120 S=KS*D/(MU-D)
130 IF S>S0 OR S<0 THEN S=S0
140 X=D*Y*(S0-S)/(D+EM*Y)
150 IF X<0 THEN X=0
160 IF X>Y*S0 THEN X=Y*S0
170 IF X>0 THEN GOTO 210
180 S2=S0
190 X2=0
200 GOTO 290
210 D2=R*D
220 A=MU*Y-D2*Y
230 B=-Y*S*MU+Y*S*D2-D2*KS*Y-MU*X
240 C=Y*S*D2*KS
250 Q=SQR(B*B-4*A*C)
260 S2=(-B-Q)/2*A
270 IF S2<0 THEN S2=0
280 X2=X+Y*(S-S2)
300 PRINT D,X,S,X2,S2
310 D=D+.1
320 IF D<MU+.2 THEN 120
330 GOTO 80
```

APPENDIX 5

BASIC Programs for Continuously Inoculated Fermenter

```
10 REM TWO-CULTURE CHEMOSTAT
20 MU=2!
30 KS=4E-03
40 S0=10
50 Y=.45
60 YZ=.42
70 MZ=1.75
80 KZ=.025
90 INPUT "INIT. CONC. OF Z ";IZ
100 PRINT "D        X                Z                S"
110 D=1E-04
120 S=KS*D/(MU-D)
130 IF S>S0 OR S<0 THEN S=S0
140 GZ=MZ*S/(KZ+S)
150 Z=D*IZ/(D-GZ)
160 IF Z<0 THEN Z=0
170 X=Y*S0-S*Y-Y*GZ*Z/(D*YZ)
180 IF X<0 THEN GOTO 230
190 PRINT D,X,Z,S
200 D=D+.05
210 IF D<MU+.8 THEN 120
215 PRINT:PRINT
220 GOTO90
230 A=-YZ*MZ+YZ*D
240 X=0
250 B=S0*YZ*D-S0*YZ*MZ-YZ*D*KZ-MZ*IZ
260 B=-B
270 C=-S0*YZ*D*KZ
280 T=B*B-4*A*C
290 S=(-B+SQR(T))/(2*A)
300 Z=IZ+YZ*(S0-S)
310 GOTO 190
```

CALCULATION OF D AT WHICH FAST CULTURE WASHES OUT

```
10 REM SCALC
15 PRINT "   A         IZ         S         D"
20 INPUT "A =";A
30 S0=5
32 MU=2!
34 KS=.1
40 YZ=.47
50 INPUT "IZ=";IZ
60 S=A*IZ/YZ+A*S0-S0
70 S=S/(A-1)
80 D=MU*S/(KS+S)
90 PRINT A,IZ,S,D
100 GOTO 50
```

APPENDIX 5B

FORTRAN Listing for Continuously Inoculated Fermenter

```
C   TWO CULTURES IN A CHEMOSTAT, ONE CONTINUOUSLY INOCULATED
        DIMENSION XPLOT(200),ZPLOT(200),YPLOT(200),SPLOT(200)
        REAL KS, MUMAX,YZ,KZ,IZ,MZ
        STEP=0.05
C   ZERO ARRAYS AND SET ABSCISSA FOR GOOD
        DO 3 I=0,200
        XPLOT(I)=0.05*I
        YPLOT(I)=0.0
        ZPLOT(I)=0.0
        SPLOT(I)=0.0
3       CONTINUE
998     CALL GRESET
5       WRITE(1,101)
        MUMAX=2.0
        KS=.1
        KZ=.11
        Y=0.5
        EM=0.00001
1       WRITE(1,107)
        READ(1,*) MZ
        WRITE(1,109)
        X=1.0
        READ(1,*) YZ
        WRITE(1,110)
        READ(1,*) SO
        WRITE(1,111)
        READ(1,*) IZ
        CALL GRESET
13      D = 0.0001
        IF(MUMAX-4.) 15,11,11
11      WRITE(1,103)
        GO TO 5
15      IF(KS.GT.2.) GO TO 11
        IF(SO.GT.101.) GO TO 11
        IF(Y.GT..99) GO TO 11
        IF(EM.GT.0.7) GO TO 11
        I=0
30      IF(X.LT.0.01) GO TO 37
        S=KS*D/(MUMAX-D)
        GZ=MZ*S/(KZ+S)
        Z=D*IZ/(D-GZ)
        IF (Z.LT.0.0) Z=0.0
        X=Y*SO-S*Y-Y*GZ*Z/(D*YZ)
        IF (X) 37,37,23
37      X=0.0
        A=D*YZ-YZ*MZ
        IF(IZ) 23,23,38
38      B=-SO*YZ*D+SO*YZ*MZ+YZ*D*KZ+MZ*IZ
        C=-SO*YZ*D*KZ
        TERM=B*B-4.0*A*C
        S=(-B+SQRT(TERM))/(2.0*A)
        Z=IZ+YZ*(SO-S)
23      IF(Z.LT.0.0) Z=0.0
        IF(S.LT.0.0) S=0.0
```

```
              IF(X.GT.SO) X=0.0
              YPLOT(I)=X
              SPLOT(I)=S
              ZPLOT(I)=Z
              D = D + STEP
              I=I+1
              IF(I-105) 30,99,99
      99      CONTINUE
              CALL ENTGRA
              CALL SET2D
              CALL INIT(-3.,-4.)
              CALL QCALE(0.,2.,8.,0)
              CALL QCALE(0.,5.,8.,1)
              CALL XLINAX('DILUTION RATE',14,8.)
              CALL YLINAX('CONCENTRATION',13,8.)
              CALL DATLIN(XPLOT,YPLOT,100,2,1,1)
              CALL DATLIN(XPLOT,SPLOT,100,1,1,1)
              CALL DATLIN(XPLOT,ZPLOT,100,3,1,1)
              CALL MOVE(2.9,2.8)
              CALL TEXT('FAST    SLOW  ',11)
              CALL MOVE(0.0,2.3)
              CALL TEXT('INIT. Z= ',9)
              CALL RMOVE(1.6,0.0)
              CALL NUMBRQ(IZ,2,3)
              CALL MOVE(0.0,2.5)
              CALL TEXT('SO =',4)
              CALL RMOVE(0.6,0.0)
              CALL NUMBRQ(SO,1,3)
              CALL MOVE(2.5,2.5)
              CALL TEXT('MU',2)
              CALL RMOVE(0.5,0.0)
              CALL NUMBRQ(MUMAX,2,3)
              CALL MOVE(3.7,2.5)
              CALL NUMBRQ(MZ,2,3)
              CALL MOVE(2.5,2.1)
              CALL TEXT('Y ',2)
              CALL RMOVE(0.5,0.0)
              CALL NUMBRQ(Y,2,3)
              CALL MOVE(3.7,2.1)
              CALL NUMBRQ(YZ,2,3)
              CALL MOVE(2.5,1.7)
              CALL TEXT('KS',2)
              CALL RMOVE(0.5,0.0)
              CALL NUMBRQ(KS,2,3)
              CALL MOVE(3.7,1.7)
              CALL NUMBRQ(KZ,3,3)
              CALL EXITGR
              WRITE(1,106)
              READ(1,112) IQ
              IF(IQ.LT.0) GO TO 999
              IF(IQ.NE.1) GO TO 1
              CALL ENTGRA
              CALL HDCOPY(ICOP)
              CALL EXITGR
```

```
        GO TO 998
101     FORMAT(12HTWO CULTURES       )
9       FORMAT(F4.1,2H I ,100A1)
102     FORMAT(5F10.4)
103     FORMAT(// 40H CRAZY SPECS, SEE INSTRUCTIONS.
106     FORMAT(// 20H STOP NOW?              )
107     FORMAT(6HMUMAX    )
108     FORMAT(3HKS    )
109     FORMAT(2HY    )
110     FORMAT(3HSO    )
111     FORMAT(3HIZ    )
112     FORMAT(I2)
999     CALL EXIT
        STOP
        END
```

APPENDIX 6

Listing or STERIL Game

```
C  STERILIZATION TEACHING GAME, H. BUNGAY, MAY 1979
      REAL NO
      IMPLICIT REAL*8(A-H,O-Z)
      DATA   YES/3HYES/, NO/3HNO /
      WRITE(3,108)
      VK = 5.0E13
      EACT = 69700.0
      R = 1.987
      VACT = 26000.0
      VZERO = 100.0
      XZERO = 6.5E12
      A = 7.94D38
   20 WRITE(1,100)
      READ(1,*) TEMP
      WRITE(1,102)
      IF(TEMP.LT.0.) GO TO 99
      READ(1,*) T
      COST = 0.0
      C = A * DEXP(-EACT/(R * (TEMP + 273.1)))
      X = XZERO * DEXP(-C*T)
      VC = VK * DEXP(-VACT/(R*(TEMP+273.)))
      V = VZERO * DEXP(-VC * T)
      IF (T - 2.0) 21,21,22
   21 WRITE(1,109)
      COST = ((TEMP - 60.0) * 350.0) + 40000.0 + 800.0 * T
      GO TO 23
   22 DIFF = TEMP - 40.0
      THETAH = DIFF/(0.0059*313. - DIFF *0.003)
      EXNO = EACT / (R * 313.0 )
      EXFACT = -37.4 + ( EXNO *(1.0+0.003*THETAH)/(1.0+0.0089*THETAH))
      DELHT = 3.68E26 * DEXP(-EXFACT)/(EXFACT * EXFACT)
      X = X * DEXP(-2.0 * DELHT)
      V = V * DEXP(-VC*4.0)
   23 COST = COST + 200000.
      WRITE(1,110) X
      IF(X - 1.0) 1,2,2
    2 WRITE(1,104)
      GO TO 20
    1 VRATIO = 100.0 * V / VZERO
      WRITE(1,111) VRATIO
      PROFIT = 1.6E6*(1.0-X)*V/(15.0+V) - COST
      WRITE(1,112) PROFIT
      GO TO 20
```

```
100 FORMAT(1X, 30HTYPE STERILIZATION TEMPERATURE         )
101 FORMAT(G15.4)
102 FORMAT( 29H TYPE TIME FOR STERILIZATION   )
103 FORMAT(2G10.3)
104 FORMAT(10X,50HYOU LOSE, ALL THE FERMENTERS WILL BE CONTAMINATED
108 FORMAT( 2X,20HSTERILIZATION  GAME     )
109 FORMAT(  '  FOR SUCH A SHORT TIME, IT IS BEST TO USE A CONTINUOU
   2 STERILIZER    ',/    ' YOUR COSTS WILL BE FIGURED ACCORDINGLY
110 FORMAT(24H ORGANISMS / FERMENTER =    , E10.3)
111 FORMAT( 18H % VITAMINS LEFT =    , F7.3 )
112 FORMAT( 19H YEARLY PROFIT = $   , F12.2)
 99 STOP
    END
```

APPENDIX 7

Listing for CHROMO Game

```
C         CHROMOTOGRAPHY GAME, H. BUNGAY 1979
C
C              FORTRAN VERSION
C              BY MARC S. PALLER
C
C         ALTERED BY JAMES GASTON
C         TO PLOT TOTAL OUTPUT
C              % PURITY
C         AND  94% PURE LINE.
C
C   MAJOR REVISION FOR AMMONIUM SULFATE PRECIP.  BUNGAY , JULY 1982
C  FIXED UP COEFFICIENTS 12/22/82
      REAL Y2(20),Y(20),Y1(20),X1(20),PURITY,DIAMET,V(20),K,S(42),
     * TOTAL, YIELD,EQUIP,FEEDST,KG,LABOR,AMSUL,PERCNT,SOL(3),LIVER(3)
     * CAKE(3),BATCH(20),B(20),BC(20),BD(20)
      DATA S(1)/1.2/,S(2)/1.2/,S(3)/1.18/,S(4)/3.1/,S(5)/9.1/
     *,S(6)/14.1/,S(7)/8.8/,S(8)/1.3/,S(9)/1.2/,S(10)/1.3/,S(11)/
     *3.0/,S(12)/9.1/,S(13)/14.9/,S(14)/9.0/,S(15)/1.1/,
     *S(16)/1.1/,S(17)/1.2/,S(18)/2.4/,S(19)/8.3/,S(20)/15.0/,
     *S(21)/5.0/,S(22)/1.2/,S(23)/1.3/,S(24)/1.4/,S(25)/4.6/,
     *S(26)/8.0/,S(27)/13.7/,S(28)/12.0/,S(29)/1.1/,
     *S(30)/1.19/,S(31)/1.28/,S(32)/3.0/,S(33)/9.3/,S(34)/14.4/,
     *S(35)/5.2/,S(36)/1.18/,S(37)/1.21/,S(38)/1.25/,
     *S(39)/4.9/,S(40)/9.0/,S(41)/15.2/,S(42)/15.1/
      WRITE(1,1)
1     FORMAT('1',/,/,/,10X,'CHROMATOGRAPHY SCALE-UP GAME',
     */,'YOU MUST ACCUMULATE CRUDE ENZYME FROM AMMONIUM SULFATE STEPS'
     */,'LIVER COSTS $2.48/KG UP TO 100, THEN $2.14',/,
     *'LABOR COST DEPENDS ON BATCH SIZE',/,'ONE PERSON FOR <100KG',
     */,'TWO PERSONS FOR <850, 3 FOR V. LARGE BATCHES.')
C
C    INITIALIZING SPECIFIC VARIABLES
C
C SUBSCRIPT V( ) FOR EASE OF HANDLING ?
C B(I) FOR ENZYME STOCKPILE, BC( ), BD( ) FOR IMPURITIES
      DO 31 I=1,20
      C=0.0
      Y(I)=0.0
      Y1(I)=0.0
      B(I)=0.0
      BC(I)=0.0
      BD(I)=0.0
      X1(I)=0.0
31    CONTINUE
      IFLAG=0
      WRITE(1,1241)
```

```
1241   FORMAT(/,'YOU MAY ENTER CRUDE BATCHES FROM PREVIOUS RUNS.',
      */,'TYPE 0 TO OMIT OR QUIT.',/,'ENTER BATCH NO:',/,
      *'MG. OF ENZYME:',/,'PURITY:')
       READ(1,*) N
       IF (N.LT.1) GO TO 1108
       READ(1,*) B(N)
       READ(1,*) PURITY
       BD(N)=(B(N)-PURITY*B(N))/(3.0*PURITY)
       BC(N)=2.0*BD(N)
       GO TO 31
1108   I=C/4.0
       S1=499.0-ABS(0.5*(65.0-C)*(65.0-C))
C    THIS CALC THE SOLUBILITIES OF THREE MAJOR COMPONENTS
       S2=138.0-ABS(0.12*(60.0-C)*(60.0-C))
       S3=401.0-0.21*(80.0-C)*(80.0-C)
       IF(S1-30.) 1102,1102,1103
C  A SMALL AMOUNT OF THE IMPURITIES ALWAYS PPTS., GOODIES  CAN=0
1102   S1=30.1
1103   IF(S2) 1104,1105,1105
1104   S2=0.0
1105   IF(S3-15.) 1106,1106,1107
1106   S3=15.123
1107   CONTINUE
       X1(I)=C
       Y1(I)=S2/139.0
       Y2(I)=S2/(S1+S2+S3)
       C=C+4.0
       IF (C-81.0) 1108,1101,1101
C   PLOT YIELD AND PURITY OF ENZYME
1101   CALL GRESET
       CALL ENTGRA
       CALL SET2D
       CALL INIT(-1.0,-1.0)
       CALL QCALE(0.0,75.0,6.0,0)
       CALL QCALE(0.0,0.9,4.0,1)
       CALL XLINAX('PER CENT AMMONIUM SULFATE',25,6.0)
       CALL YLINAX('YIELD , PURITY',15,4.0)
       CALL DATLIN(X1,Y1,20,2,1,1)
       CALL DATLIN(X1,Y2,20,4,1,1)
       CALL EXITGR
       JFLAG=0
C  SET V() TO V. SMALL NO.
       DO 2 I=14,20
       V(I)=1.0E-05
2      CONTINUE
1269   WRITE(1,3)
3      FORMAT(/,2X,'KG. OF LIVER TO BE PROCESSED ?')
       READ(1,*)K
4      FORMAT(F10.2)
       WRITE(1,1210)
1210   FORMAT(/,2X,'PER. CENT AMMONIUM SULFATE TO ADD ?')
       READ(1,*) C
       CALL GRESET
       S1=490.0-ABS(0.5*(65.0-C)*(65.0-C))
       IF (S1.LT.0.0)    S1=2.21
       S2=138.8-ABS(0.12*(60.0-C)*(60.0-C))
       IF (S2.LT.0.0)    S2=0.0
       S3=401.1-0.21*(80.0-C)*(80.0-C)
```

```
              IF (S3.LT.0.0)     S3=9.123
              YIELD=S2*K
              CL=902.38
C  CL IS COST OF LABOR
              IF (K.LT.100.0)  CL=475.05
              IF (K.GT.850.0)  CL=1409.82
              CG=2.38*K
              IF (K.LT.100.0)  CG=2.48*K
              IF (K.GT.850.0)  CG=2.14*K
C  CA IS COST OF AMMONIUM SLFATE
              CA=0.00283*C*K
              C4=CG+CA+CL
C  COSTS ARE FUNCTIONS OF SCALE, LARGER AMTS OF LIVER COST LESS
              WRITE(1,1400) CG,CA,CL,C4
1400          FORMAT(/,20X,'COSTS',/,2X,'LIVER',12X,F10.2,/,2X,'AMMONIUM SULFA
             * ',F10.2,/,2X,'LABOR',12X,F10.2,/,2X,'TOTAL',10X,F12.2)
              V(20)=V(20)+C4
              P=100.0*S2/138.1
              PURITY=S2/(S1+S2+S3)
              WRITE(1,1220) YIELD,P,PURITY
1220          FORMAT(/,2X,'YIELD = ',F12.1,' MG.  OR ',F6.2,' PER CENT',
             */,2X,'PURITY = ',F5.4)
C
C     ADJUST AMOUNTS OF MAJOR COMPONENTS
              N=1
1130          IF (B(N)-.9) 1140,1140,1150
1150          N=N+1
              IF (N-20) 1130,1130,1160
1160          WRITE(1,1230)
1230          FORMAT(/,2X,'TOO MANY BATCHES.  WILL BE POOLED WITH BATCH 20.')
              B(20)=B(20)+K*S2
              BC(20)=BC(20)+K*S1
              BD(20)=BD(20)+K*S3
              GO TO 1170
1140          WRITE(1,1240) N
1240          FORMAT(/,2X,'THIS IS ASSIGNED AS BATCH ',I2)
              B(N)=K*S2
              BC(N)=K*S1
              BD(N)=K*S3
1170          S1=0.0
              S2=0.0
              S3=0.0
              T1=0.0
              T2=0.0
              T3=0.0
1271          WRITE(1,1250)
1250          FORMAT(/,2X,'YOU NOW HAVE THE FOLLOWING BATCHES :')
              DO 1180 N=1 ,20
                IF(B(N)-0.9) 1180,1180,1171
1171          PURITY=B(N)/(B(N)+BC(N)+BD(N))
              WRITE(1,1260) N,B(N),PURITY
1260          FORMAT(/,2X,'BATCH ',I2,'   MG. = ',F10.1,'   PURITY = ',F5.4)
1180          CONTINUE
C
              WRITE(1,603)
603           FORMAT(/,/,2X,'DO YOU WISH TO TERMINATE THIS SESSION'
             *,/,50X,' (YES OR NO)')
              JFLAG=0
```

```
              READ(1,111) ANS
111           FORMAT(A3)
              IF(ANS.EQ.'YES')GO TO 1000
              WRITE(1,1264)
1264          FORMAT(/,2X,'TYPE A MINUS NO. TO MAKE MORE BATCHES, TYPE 0 T
             *',/,2X,'CONTINUE TO CHROMATOGRAPHY, OR TWO NO. TO POOL BATCHE
              READ(1,1263) N
              IF (N) 1101,1267,1268
1268          READ(1,*) N1
              B(N)=B(N)+B(N1)
              BC(N)=BC(N)+BC(N1)
              BD(N)=BD(N)+BD(N1)
              B(N1)=0.0
              GO TO 1271
1267          WRITE(1,1262)
1262          FORMAT(/,2X,'TYPE BATCH NO. FOR CHROMATOGRAPHY')
              READ(1,1263) N
              IF(N.GT.20) GOTO 1271
              IF(B(N).LT.1.0) GOTO 1271
1263          FORMAT(I2)
332           WRITE(1,5)
5             FORMAT(/,2X,'ENTER MG. OF ENZYME APPLIED TO COLUMN (F FORMAT
              READ(1,*)V(9)
              IF(B(N)-V(9)) 340,341,341
340           WRITE(1,1265)
1265          FORMAT('INSUFFICIENT ENZYME IN THIS BATCH.')
              GO TO 332
341           CONTINUE
              V(11)=V(9)*BC(N)/B(N)
              V(12)=V(9)
              P=B(N)/(B(N)+BC(N)+BD(N))
              V(13)=V(9)*BD(N)/B(N)
              V(7)=0.001*V(9)/P
              B(N)=B(N)-V(9)
              BC(N)=BC(N)-V(11)
              BD(N)=BD(N)-V(13)
              WRITE(1,7)V(7)
7             FORMAT(/,/,11X,'THIS IS ',F7.3,' GRAMS OF PROTEIN.')
100           WRITE(1,8)
8             FORMAT(/,2X,'ENTER THE DIAMETER OF THE COLUMN (F FORMAT) IN
              READ(1,*)DIAMET
C
C     GRAMS PER LITER OF RESIN
C
              C=V(7)/(0.05*DIAMET*DIAMET)
              WRITE(1,10)C
10            FORMAT(/,/,11X,'THE LOADING IS ',F9.4,' GRAMS/LITER OF PROT
             *',/,/,11X,'DO YOU WISH TO USE THIS LOADING (YES OR NO)')
              READ(1,11)ANS
11            FORMAT(A3)
              IF(ANS.EQ.'NO')GO TO 100
              WRITE(1,1242)
1242          FORMAT(/,/,'ADSORBENT PRICE DOES NOT RELATE TO PERFORMANCE'
             */,'BUT PRICES PER L ARE:',/,2X,'ADSORBENT    PRICE')
```

```
          DO 41 N=1,6
          CRESIN=N
          CRESIN=104.13*SQRT(CRESIN)
          WRITE(1,1243) N,CRESIN
 1243     FORMAT(4X,I1,4X,'$',F7.2)
 41       CONTINUE
          WRITE(1,12)
 12       FORMAT(/,/,'SELECT ADSORBENT FROM 1 TO 6 (I FORMAT)')
          READ(1,13) NADSOR
 13       FORMAT(I1)
C      IF ADSORBANT IS LESS THAN 1 IT IS SET EQUAL TO 1
C      IF ADSORBANT IS GREATER THAN 6 IT IS SET EQUAL TO 6
C
C
          IF(NABSOR.GT.6)NADSOR=6
          M=1+7*(NADSOR-1)
          IF(NADSOR.LE.1)M=1
C
C      DUMMY DATA ASSIGNED THROUGH DATA STATEMENT
C
          TEMP4=S(M+6)-C
          IF(TEMP4.GT.0.0)GO TO 300
          IFLAG=1
          V(8)=1.2-TEMP4
          GO TO 200
 300      V(8)=1.0+0.06*DIAMET/(TEMP4+0.7)
C     ABOVE CORRECTS FOR DIAM AND LOADING, 0.7 PREVENTS SMALL DIVIDE
C
C     ADJUST FOR CAPACITY AND DIAMETER
C
 200      DO 21 I=1,3
          S(I+M-1)=S(I+M-1)*V(8)
 21       CONTINUE
          V(4)=0.0
          V(10)=0.0
C
C     SET X, FRACTION NUMBER EQUAL TO ZERO
C LL IS FLOAT FOR INDEXING AND TO SET ABSCISSA FOR PLOT
C
          X=0.0
 650      X=X+1.0
          IF(X.GT.20.0)GO TO 400
          LL=IFIX(X)
          X1(LL)=FLOAT(LL)
C
C      NORMAL DISTRIBUTION EQUATION
C
          F=20.0+2.0+SQRT(0.1*C)
          DO 22 I=1,3
          A=0.5*(X-S(M+2+I))*(X-S(M+2+I))/(S(M+I-1)*S(M+I-1))
          V(I)=0.001*F*V(I+10)*EXP(-A/S(M+I-1))
 22       CONTINUE
C
C     V(4) IS THE ENZYME, V(5) IS THE TOTAL STUFF, AND V(10) IS PRODUCT
```

```
      C
                V(4)=V(4)+V(2)
                V(5)=V(1)+V(2)+V(3)+1.0E-06
                N=0
                T1=T1+V(1)
                T2=T2+V(2)
                T3=T3+V(3)
                Y(LL)=V(1)+V(2)+V(3)
                Y1(LL)=V(2)/V(5)
                IF(V(2)/V(5).LT.0.94)GO TO 500
                V(10)=V(10)+V(2)
                GO TO 650
      500       IF(V(2).LT.0.25)GO TO 650
      C
      C         ACCUMULATE ENZYME IN BAD FRACTION.
      C
                IF(V(2)/V(5).LT.0.4)GO TO 650
                S1=S1+V(1)
                S2=S2+V(2)
                S3=S3+V(3)
                GO TO 650
      C
      C         Y IS TOTAL OF 1,2,3
      C
      C         Y1 IS THE % PURITY
      C         Y2 IS THE 94%C
      C
      C
      400       CONTINUE
                CM11=AMAX1(Y(1),Y(2),Y(3),Y(4),Y(5),Y(6),Y(8),Y(9),Y(10),
               *Y(11),Y(12),Y(13),Y(14),Y(15),Y(16),Y(17),Y(18),Y(19),Y(20))
                CM21=AMAX1(Y1(1),Y1(2),Y1(3),Y1(4),Y1(5),Y1(6),Y1(7),Y1(8),
               *Y1(9),Y1(10),Y1(11),Y1(12),Y1(13),Y1(14),Y1(15),Y1(16),
               *Y1(17),Y1(18),Y1(19),Y1(20))
                CM3=1.0
                DO 101 I=1,20
                Y2(I)=.94*CM3
                Y(I)=Y(I)/CM11
      101       CONTINUE
                CALL GRESET
                CALL ENTGRA
                CALL SET2D
                CALL INIT(-4.5,-1.0)
                CALL QCALE(1.,20.0,4.5,0)
                CALL QCALE(0.,CM3,5.0,1)
                CALL XLINAX('FRACTION NUMBER',15,4.5)
                CALL YLINAX('PER CENT',8,5.0)
                JKL=20
                CALL DATLIN(X1,Y,JKL,2,1,1)
                CALL DATLIN(X1,Y2,JKL,2,1,1)
                CALL DATLIN(X1,Y1,JKL,4,1,1)
      C         CALL HDCOPY(ICOP)
                CALL EXITGR
                DO 39 I=1,20
```

```
              Y1(I)=0.0
              Y(I)=0.0
    39        CONTINUE
              IF(IFLAG.EQ.0)GO TO 262
              WRITE(1,20)
    20        FORMAT(/,/,21X,10('*'),' COLUMN IS OVERLOADED WITH PROTEIN',
             *1X,10('*'))
              IFLAG=0
    262       YIELD=V(9)*V(10)/V(4)
              YIELD1=100.0*(V(10)/V(4))
              WRITE(1,250)YIELD,YIELD1
    250       FORMAT(/,/,/,/,/,/,/,/,47X,'YIELD = ',F7.3,' MG..',3X,F7.3,
             *1X,'PER CENT')
              V(19)=V(19)+YIELD
              C2=400.0+55.0*DIAMET
              CRESIN=NADSOR
              C3=DIAMET*DIAMET*30.1*SQRT(CRESIN)
              C5=C2+C3
              V(17)=V(14)+V(15)+V(16)
              V(18)=V(18)+V(15)*V(9)/V(4)
              S1=S1*V(11)/T1
              S2=S2*V(12)/T2
              S3=S3*V(13)/T3
              WRITE(1,1405)
    1405      FORMAT(/,45X,'OVERALL COSTS INCLUDE ALL LIVER PROCESSING.')
              WRITE(1,251)C2,C3,C5
    251       FORMAT(/,/,47X,'COSTS : $ ',F9.3,' FOR LABOR AND ASSAYS',
             */,57X,F9.3,' FOR MATERIALS',
             */,47X,'TOTAL    $',1X,F9.3)
              V(20)=V(20)+C5
              IF(YIELD.LT.0.01)GO TO 600
              TEMP7=C5/YIELD
              TEMP8=V(20)/V(19)
              WRITE(1,260)TEMP7,V(19),TEMP8
    260       FORMAT(/,/,47X,'PER MG. COST IS $ ',F9.3,
             */,44X,'TOTAL MG. COLLECTED = ',F9.3,' AT $ ',F9.3,' PER MG.')
    600       TEMP9=V(15)*100.0/V(17)
              WRITE(1,602)
    602       FORMAT(/,43X,'SIDE FRACTIONS OF REASONABLE PURITY ARE POOLED.')
              K=1.0
              WRITE(1,1246)
    1246      FORMAT(/,/,/,43X,'TYPE ANY NUMBER TO MOVE ON.')
              READ(1,*) N
              N=1
              CALL GRESET
              GO TO 1130
    1000      WRITE(1,1225)
    1225      FORMAT(/,'WRITE DOWN BATCH INFO IF YOU PLAN TO RETURN.')
              STOP
              END
```